Classified by the Israeli Air Force until now, this rare group
photograph of all eight mission pilots was shot just minutes before
takeoff. They are in bombing order (*row by row from the front, from
right to left*): Zeev Raz, Amos Yadlin; Doobi Yaffe, Hagai Katz;
Amir Nachumi, Iftach Spector; Relik Shafir, Ilan Ramon.

RAID ON THE SUN

RODGER W. CLAIRE

INSIDE ISRAEL'S SECRET CAMPAIGN THAT DENIED SADDAM THE BOMB

BROADWAY BOOKS   NEW YORK

PRINTED IN THE UNITED STATES OF AMERICA

BROADWAY BOOKS and its logo, a letter B bisected on the diagonal,
are trademarks of Random House, Inc.

Visit our website at www.broadwaybooks.com

First edition published 2004

Book design by Laurie Jewell

Library of Congress Cataloging-in-Publication Data
Claire, Rodger William.
Raid on the sun : inside Israel's secret campaign that denied Saddam
the bomb / Rodger W. Claire.— 1st ed.
p.   cm.
1. Arab-Israeli conflict—Iraq. 2. Osirak Nuclear Reactor Bombing,
Iraq, 1981. 3. Nuclear reactors—Iraq. I. Title.

DS119.7.C574 2004
956.7044'1—dc22
2003055887

Photo credits, in order of appearance: Osirak mission pilots © Israel Defense Forces;
Saddam Hussein and Jacques Chirac © AP/Wide World; David Ivry © David
Rubinger/Corbis; Yehoshua Saguy © AP/Wide World; Menachem Begin and Raphael
Eitan © AP/Wide World; Yigel Yadin and F-16 © David Rubinger/Corbis; F-16 in flight
© Aero Graphics, Inc./Corbis; Ilan Ramon © Reuters NewMedia Inc./Corbis; desolated
site of Osirak © AFP/Corbis

ISBN 0-7679-1400-7

1   3   5   7   9   10   8   6   4   2

To Grace,

my mother and the first scrivener

And, naturally, to the three muses:

Ann, Wren & Kelsey

CONTENTS

AUTHOR'S NOTE

Avi, my driver, had gone suddenly ashen. His eyes, usually mischievously bright, were hard and furtive.

"This is no good," he said.

We had stopped at a turnout on the summit of Hebrew University in Jerusalem, just across from the Frank Sinatra Student Center, so I could take in the dramatic view of the Old City, the shiny golden cupola of the Dome of the Rock sparkling among the whitewashed turrets and buildings below. A dusty, beat-up van had pulled in behind us and several young Arab teens jumped out. Avi immediately grew tense.

"We should go!" he said again. "This is no good."

"Come on," I said playfully, trying to get a rise out of him.

But Avi was not biting.

"No good," he said.

We climbed into his taxi, and as we sped off, I looked out the back window. Half a dozen Palestinian youths were standing on the asphalt, watching us leave. Avi's reaction—or overreaction, I thought—depressed me. It showed how deeply the distrust and fear between Israelis and Palestinians had burrowed as the *intifada* dragged into its second year the summer of 2002. For two weeks Avi had been chaperoning me up and down Israel in his taxi as I in-

terviewed the Israeli Air Force (IAF) pilots who took part in the infamous bombing of Iraq's Osirak nuclear reactor in 1981. This day
we still had to drive to a remote village above Ramat Hod Sharon,
north of Tel Aviv, and Avi was nervous about taking the main road
that cut through "the Territories," the Palestinian-controlled West
Bank, now flooded with Israeli army troops trying to stop the infil-
tration of suicide bombers across the Green Line. We ended up
taking the long way around, adding another half hour to the drive. I
thought Avi was being paranoid.

A week later a Palestinian boy exploded a remote-control bomb
during lunchtime inside the very same Frank Sinatra Center, killing seven students, including five Americans, and injuring scores
of others. A week after that an Israeli driver and his wife were
shot to death on the "shortcut" along the West Bank that Avi had
refused to take.

By then, however, I was back home, safe and sound in seaside
Santa Monica, California, which, with the exception of dozens of
northern Italian restaurants instead of kebab grills and the absence
of the occasional suicide bomber, is very much like Tel Aviv. But I
had brought something home with me, a valuable lesson that
would help me in the writing of this book and that no amount of
interviews and research could ever teach: what it is like to live constantly at risk.

Like most Americans, I first learned of Israel's attack on an Iraqi
nuclear reactor through newspaper accounts. At the time, June
1981, the attack seemed rather provocative, even "reckless," to
use the term employed by Secretary of State Alexander Haig—
especially given the simmering tensions of the Middle East and
the delicate Egypt-Israeli truce in the wake of Camp David. Iraq
was one of those faraway Arab countries that seemed vaguely hos-

tile, like Yemen or Syria, but one that in recent years had become an increasingly important U.S. trading partner in the region. But it remained relatively unknown. Next-door neighbor Iran and the ayatollah dominated the evening news back then. Few people had even heard of Saddam Hussein, let alone his weapons of mass destruction.

It wasn't until some four years later, when I was working as an editor at *Los Angeles* magazine, that I began to understand the enormous consequences of the Israeli air raid on the nuclear complex in al-Tuwaitha outside of Baghdad. It came one day after a contact in Southern California's then-burgeoning defense industry, who had been briefed on the classified raid, related to me—off the record—the inside story of the mission: that Saddam Hussein had a secret atomic-weapons program and planned to use the French-built Osirak reactor to produce weapons-grade plutonium; that the targeting of the reactor by the Israeli pilots was one of the most accurate bombing missions in modern warfare; that the F-16s the Israelis flew were somehow made to fly far beyond the envelope of their design specs; and that the pilots flew six hundred miles no more than a hundred feet off the ground.

I thought at the time, Geez, what a great book that would make! Except for one problem: the mission and even the names of the pilots in the raid had all been put under wraps by the Israel Defense Forces (IDF).

So it remained until June 2001, when I spotted in the *Los Angeles Times* a short interview with Israel's ambassador to the United States, Gen. David Ivry, the IAF commander who had originally planned the raid on Osirak exactly twenty years earlier. Guessing that Israel might finally be more inclined to open up about the mission—given the perception of Saddam Hussein in the years since the 1991 Persian Gulf War, the discovery of extensive biological, chemical, and nuclear weapons in Iraq, and

Hussein's refusal in 1998 to allow U.N. weapons inspectors back into his country—I sent a letter to the ambassador to request an interview.

By the first week in September 2001, I was meeting with Ambassador Ivry on the second floor of Israel's striking mansion compound, situated just down the street from the vacant-looking Ethiopian embassy along the mini–Embassy Row in Washington's leafy, redbrick Van Ness district. We talked all morning and again all the next morning. By the time I left, I had a rundown of the entire history of the action. And, thanks to the embassy's military attaché, Brig. Gen. Rani Falk—who, in a remarkable piece of luck, turned out to be one of the original group of pilots who had trained for the secret bombing attack—I had the name and telephone number of the squadron leader, Zeev Raz. I also had General Falk's assurance that, depending on the individual decision of each pilot, I could meet with every member of the team—in Israel. I would be the only journalist in twenty years to learn the names of and meet face-to-face with all eight Israeli pilots who had flown to Baghdad in 1981.

Excited and exhilarated, I returned to my Washington, D.C., hotel room to make plans to fly home early. My return ticket to Los Angeles International Airport was booked for Tuesday morning. That was four days away, and I had already wrapped up my business. But as it turned out, my round-trip ticket on American Airlines from Dulles International to L.A. could not be changed. I had purchased a specially discounted seat, and as part of the agreement I had to stay in Washington through the weekend—obviously to subsidize the hotel industry. The earliest I could book a flight back to Los Angeles was Flight 77, 9:00 A.M., Monday, September 10, 2001. I booked it and had a wonderful return trip home—the flight attendant not only gave me free earphones to watch the in-flight movie but also an extra cookie with lunch.

The following morning I awoke in Los Angeles at 5:30 A.M., still on East Coast time. I turned on one of the early-morning talk shows while my youngest daughter dressed for her second day as a freshman in high school. Out of the corner of my eye I noticed that the broadcast had cut away to a special breaking story: on the screen I saw a distant shot of what looked like a small plane, maybe a Learjet, crashing into the steel-and-glass side of one of the Twin Towers in New York City. An hour later, to the disbelief of all of America, it was clear what had happened. Next came reports of the hijacked Flight 77 out of Dulles, which had circled for an hour before slamming into the Pentagon, killing all aboard—including, I realized with horror, the flight crew I had flown with the day before, including my wonderfully generous attendant. It was chilling to know I could have easily been on that flight.

The world, at least the world of Americans, had changed immutably within hours. And so too had the world of the book I was planning to write about the Israeli raid on Osirak. No longer was my proposed book simply a great military tale. Given the increasing malignancy of Islamic fanaticism, the war on terror, and, finally, the annunciation of the so-called Bush Doctrine, which held that the United States was justified in attacking peremptorily any enemy it considered a threat, without warning, anywhere in the world, the Israeli raid in 1981 became overnight not only relevant but perhaps a blueprint for future U.S. actions.

By the time I landed in Israel in the summer of 2002, all eight pilots who had flown the bombing mission to Osirak had agreed to talk to me. For the first time, the pilots told a reporter their personal stories of how they entered the air force, how they were chosen and trained, how it felt to fly into what IDF intelligence had characterized as a "hornet's next of AAA and SAM batteries," in which at least a quarter of the pilots were expected to be lost. Gen. Amos Yadlin even showed me actual video footage of the air raid taken from the nose cameras of the F-16s. It looked like being caught in the middle of a Fourth of July fireworks display, with the sound of exploding AAA (antiaircraft artillery) and tracers drowned out by the increasingly frenetic radio chatter coming from panicked antiaircraft gunners below. I could hear a pilot's rhythmic breathing suddenly quicken over the radio as his plane nosed down into the chalky streaks of missile contrails and hot tracer bullets.

As it turned out, the eighth pilot, Ilan Ramon, was back in Houston, Texas, waiting to board the next Columbia shuttle as Israel's first astronaut. I spoke to Ilan that summer, and we agreed to get together for a more extensive follow-up interview after he returned from the NASA shuttle's science mission the following February. Tragically, we never kept that meeting. As I watched in horror, Ramon perished when Columbia broke up over Texas on February 1, just minutes from completing its historic mission. The youngest and maybe the most beloved among the Osirak pilots, Ramon was still full of boyish energy and a self-deprecating warmth. What he was most concerned about regarding our interview was not that he be given credit for his part in the Osirak raid, but that the release of his name might expose his family to danger from Saddam Hussein. His extensive travel and exposure abroad as Israel's first astronaut would make him an easy target for Iraq's murderous Mukhabarat security agents. If the Iraqi dictator were crazy enough to attempt to assassinate President George H. Bush, he was easily mad enough to want to liquidate one of the men responsible for ending his nuclear dreams.

Indeed, the constant dread of attack by Saddam Hussein that so colored Israel's wrenching decision to take out his nuclear reactor twenty years earlier was still palpable in all the pilots I met. It was one of the reasons why the IDF insisted that the pilots' names remain classified for two decades. Many of the team had gone on to second careers in electronics or Israel's defense industry and traveled abroad extensively. None of them wanted to be surprised by an Iraqi bullet on a street in Istanbul or New Delhi.

Who at the time could have predicted that within a year, Saddam Hussein and his Ba'thist regime would be no more? Perhaps because of this New World Order, or because after a year of telephone and e-mail exchanges, of questions and answers and just plain talk, everyone involved in telling about the raid had come to trust one another. Or maybe, after twenty years, it was just time for the full story to be told. Or maybe for all those reasons, I was able for the first time to tell the entire story of this remarkable raid using all the real names and actual documents—from the first Israeli intelligence reports of a meeting between French prime minister Jacques Chirac and Saddam Hussein in Baghdad in 1974 to the epic political battles within Israeli prime minister Menachem Begin's government over whether to attack Iraq to the personal tales of the bombing as seen through the eyes of the pilots themselves.

It is a remarkable story of courage and conviction—and of an action that proved to be a turning point in the history of Saddam's Iraq. In fact, it could well be argued, the Coalition's stunning military victory in finally liberating Iraq from the tyranny of Saddam Hussein in May 2003 began on a cloudless evening over al-Tuwaitha on June 7, 1981.

RODGER W. CLAIRE
Los Angeles, California

CAST OF CHARACTERS

JUNE 1981
ISRAELI GOVERNMENT

ISRAEL DEFENSE FORCES ——— **Prime Minister** ——— **MOSSAD**

Menachem Begin
(pro)

*(Institute for Intelligence
and Special Operations)*

Chief of Staff
Gen. Rafael "Raful" Eitan
(pro)

Deputy Prime Minister
Yigael Yadin
(opposed)

Director General
Yitzhak Hofi
(opposed)

**Commander
Israel Air Force**
Gen. David Ivry
(pro)

Deputy Prime Minister
Simha Erlich
(pro)

Deputy Director General
Nahum Admoni
(pro)

**Chief of
Military Intelligence**
Gen. Yehoshua Saguy
(opposed)

Defense Minister
Menachem Begin
*(retained position after
resignation of Ezer
Weizman 5/80, who opposed)*
(pro)

Head of the Paris Station
David Arbel
(pro)

OSIRAK MISSION PILOTS
*(In order of bombing
position)*

Deputy Defense Minister
Mordechai Tzitori
(pro)

FIRST TEAM
No. 1 Zeev Raz
(squadron commander)
No. 2 Amos Yadlin
No. 3 Doobi Yaffe
No. 4 Hagai Katz

Agriculture Minister
Ariel Sharon
(pro)

SECOND TEAM
No. 5 Amir Nachumi
(team leader)
No. 6 Iftach Spector
No. 7 Israel "Relik" Shafir
No. 8 Ilan Ramon
*(Israel's first astronaut;
died in Columbia Space
Shuttle explosion)*

Foreign Minister
Yitzhak Shamir
(pro)

Justice Minister
Moshe Misim
(pro)

Interior Minister
Dr. Joseph Borg
(pro)

Health Minister
Eliezer Shostak
(opposed)

ABBREVIATIONS

AAA	antiaircraft artillery
AWACS	airborne warning and control system/ intelligence-gathering aircraft
BITS	built-in test systems
Com	communications
CSAR	combat search-and-rescue
GCI	ground-controlled intercept (radar)
HUD	heads-up display
IAEA	International Atomic Energy Agency (U.N.)
IP	initial point
Kfir	Israeli fighter/trainer aircraft
KH-11	top-secret photography satellite
MK-84	2,000-lb. gravity bombs
PRC	rescue/homing device
SA-7	shoulder-fired antiaircraft missile
SAM-6	Soviet-made antiaircraft missile
ZSU	mounted antiaircraft gun

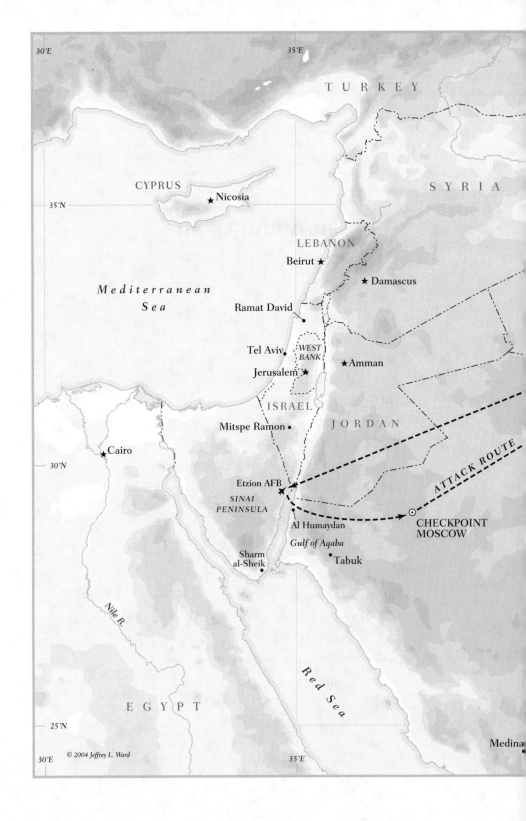

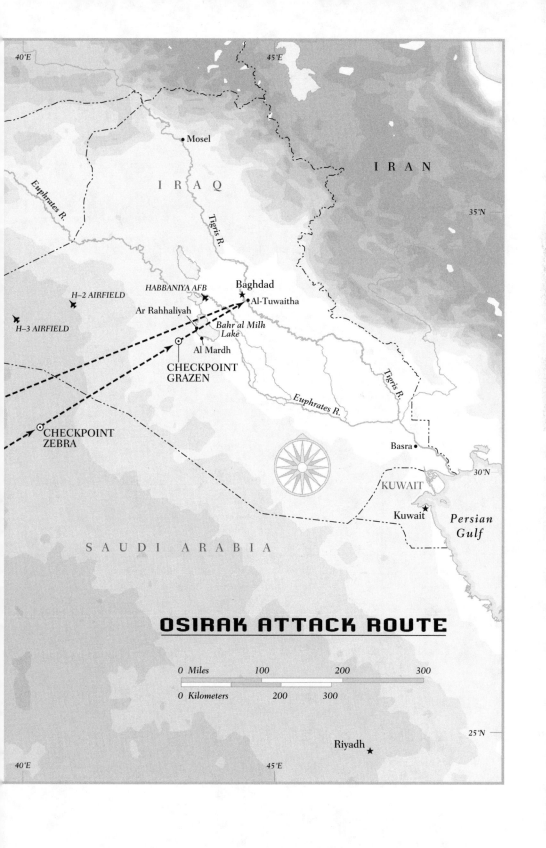

40°E

45°E

• Mosel

I R A Q

IRAN

35°N

Euphrates R.

Tigris R.

HABBANIYA AFB

Baghdad

H–2 AIRFIELD

★ Al-Tuwaitha

Ar Rahhaliyah

Bahr al Milh Lake

H–3 AIRFIELD

• Al Mardh

CHECKPOINT
GRAZEN

Euphrates R.

Tigris R.

CHECKPOINT
ZEBRA

Basra •

30°N

KUWAIT

Kuwait ★

Persian Gulf

S A U D I   A R A B I A

OSIRAK ATTACK ROUTE

0 Miles   100   200   300

0 Kilometers   200   300

25°N

Riyadh ★

40°E

45°E

RAID ON THE SUN

THE ROAD TO BABYLON

The noise of battle is in the land,
the noise of great destruction.

—JEREMIAH 50:22

JUNE 6, 1981

RAMAT DAVID, JEZREEL VALLEY, ISRAEL

General David Ivry's wife, Ofera, had invited friends in for the weekend. He had tried to make the best of it, yet over dinner that night and later over coffee he had been poor company. He was distracted, preoccupied. The guests assumed it was his "job." After all, the commander of the Israeli Air Force was bound to bring home the cares of office some days, even on the Sabbath. His wife wasn't so sure. Her husband had seemed tense ever since returning that morning from the official ceremony in Naples celebrating the change of command of the United States' 6th Fleet, stationed in the Mediterranean.

Short and compact, Ivry had the keen eyes and efficient movements of a bantamweight. Even sitting at the dinner table, he was as square and straight as an executive in a boardroom. His face, framed by short graying hair and a military mustache, was still youthful, showing none of the signs of a lifetime of holding the buck. Born in the small town of Gedera in southern Israel in 1935, Ivry was barely fourteen when the country won its independence. Like most of the men who made up the leadership of the nation, he was part of that first generation to be born, raised, and educated an Israeli—the first time a Jew could call himself that in two thousand years. And like his peers, he had been forced to defend that privilege most of his life. At eighteen he joined the IAF, and by forty had already flown combat missions in the '56 Sinai campaign, the Six-Day War, the Yom Kippur War, and the War of Attrition in '70. Though years in the military made him short on get-to-know-you conversation, he was not without warmth. He had a quick smile, and his eyes grew soft and shiny whenever he talked about his wife—or his pilots. But this evening, Ofera knew, something was wrong.

Later that night, after the guests retired and as Ivry and Ofera prepared for bed, he spoke up in spite of himself.

"Tomorrow, at sunset," the general began, looking at his wife carefully, "we will launch fighter planes to Iraq to attack a nuclear facility Saddam Hussein is using to make atomic bombs. It is a very risky mission. Never before in history has anyone bombed a nuclear reactor. If it fails, Iraq could attack us. The world may turn against us. Israel could be isolated."

Ofera did not respond at first. They had been married a long time, through four wars. He had left her to fight in each one, and each time he had left without telling her the details of his mission, leaving her to wonder and worry. And each time he had returned safely. Why should this time be any different?

"Is there nothing else to do?" she said, knowing the answer already.

"No."

Ivry hugged his wife. He had frightened her, but, almost perversely, he felt suddenly relieved. He had not slept an entire night through for weeks. But that evening, the minute his head hit the pillow, he was fast asleep.

His wife sat up awake the entire night.

Ivry got up early Sunday, drank a little coffee and picked at a roll, then kissed Ofera good-bye.

"Shalom," he said.

"Shalom."

A staff car drove the general to the small air force base in the north of Tel Aviv, in the shadow of the Mediterranean city's towering landmark utility "chimney" winking its red aircraft-warning lights. Saluting the two corporals at the main gate, Ivry's driver passed the concrete barriers and followed the narrow asphalt road to IAF headquarters, where the general had spent the last year planning this mission. Waiting for him was Maj. Gen. Yehoshua Saguy, chief of Israel's military intelligence, known as AMAN. The two generals walked to the airfield tarmac and climbed aboard a waiting Sikorsky CH-53, the helicopter's long, razor-sharp blades whipping the air above its desert-brown-camouflage fuselage. The chopper would fly the commanders the relatively short hop south to Etzion Air Force Base in what had been Egypt's Sinai Peninsula before the Six-Day War. Located inland from Eilat, the chic Israeli resort town on the Red Sea favored by Scandinavian, German, and local tourists, Etzion was part of the territory Israel was about to cede back to Egypt as part of the Camp David Accords that Menachem Begin and Anwar Sadat had signed in 1978. What effect would the surprise attack have on Sadat and these negotiations? Ivry wondered, quickly pushing the thought aside.

The CH-53 soared above the arid plains of the Negev and

southern Israel. The men spoke little during the hour-long journey. The air outside was already growing warm. Thank goodness it was still June, Ivry thought. In a month, temperatures in the Negev would be scorching. Finally, the helicopter banked left toward the gray ribbon of Etzion Airfield and the cluster of residential houses surrounding the base. It was midmorning, but the streets below were eerily quiet. As Ivry knew, most of the inhabitants had been evacuated in the preceding days—nonessential personnel and base staff had been given leave or temporary reassignment. Weekend passes had been canceled and all military personnel were confined to base. Telephone communication into and out of Etzion was cut. Few inside the base had noticed that twelve F-16 Fighting Falcons had been landing on the far runway since early Friday. Operation Babylon, code-named after the ancient biblical name for Iraq and planned in complete secrecy for more than two years, was minus six hours and counting.

The Sikorsky landed with a thud. The generals jumped from the gangway, ducking beneath the blades and holding their caps against the rotor wash as they dashed across the tarmac toward the briefing room. Walking past the camouflaged underground hangars, Ivry could see dozens of crew chiefs and maintenance techs who were readying the huge fighters. The planes below stood menacingly anonymous, tinted in brown desert camouflage, the signature blue six-point Star of David on their tails painted over for this mission. Forklifts flanked by ordnance specialists on either side ferried two-thousand-pound MK-84 bombs to the planes, where they were raised to the release clips beneath the wings of the F-16s and mounted, the ordnance techs couldn't help thinking, perilously close to the pair of external fuel tanks that also hung beneath the wings on either side of the fuselage.

As Ivry walked up the short wooden ramp to the pilots' briefing room, he was surprised to see "Raful," Gen. Rafael Eitan, chief of staff of the Israel Defense Forces. Eitan was a larger-than-life character whose exploits as the tough commander of Israel's crack paratroopers during the bloody Sinai campaigns in '67 and '73 were legendary in the IDF. With thick shoulders, a handsome, open face, and big, burly eyebrows, he looked more like a back-alley brawler than a three-star general.

Though he had suspected Eitan would come, Ivry was surprised nonetheless to see him standing there, his uniform immaculate and trim as always, but his usually animated face gaunt, his eyes ringed and tired. Raful's son, Yoram, a young IAF fighter pilot, had been killed just four days earlier right there on the base. Impetuous, irrepressibly energetic, the young pilot had lost control of his Kfir fighter during a training exercise and plummeted helplessly five thousand feet to the desert floor in a "dead man's" spin. They had interrupted the general in the middle of a mission readiness meeting to tell him of his son's death. "Raful" had left the base immediately to sit shivah, the traditional Jewish mourning period of seven days of seclusion. That was in Tel Aviv on Wednesday. Now, Sunday morning, without advising anyone, the chief of staff had requisitioned a plane and flown down by himself in order to be with the men as they began their mission, gathering now inside the briefing room for the final run-through.

Eitan caught Ivry's look of concern. He smiled wanly.

"We ask a lot of these boys, don't we?" Eitan said.

Ivry understood the question. He had also lost a son in the service several years before.

"Maybe a little more this time," Ivry replied.

Eight pilots would have to fly the new, computerized, and highly sophisticated, almost futuristic American-made F-16 Falcons nearly six hundred miles over hostile territory to bomb Iraq's nearly completed Osirak reactor in al-Tuwaitha, a heavily defended nuclear installation twelve miles south of Baghdad. The mission would be the longest, most dangerous, most technologically challenging military operation in Israel's history. It would be the first time Israeli pilots had engaged an enemy at such a distance and so far from Israel's borders. The first time sleek and speedy F-16s would attempt takeoff carrying a weight that exceeded nearly twice the planes' design specs. The first time anyone, anywhere, had bombed a nuclear reactor.

The very idea seemed somehow blasphemous. Since the beginning of the Manhattan Project in the forties, statesmen, philosophers, even the physicists themselves had questioned the hubris of attempting to harness the frightening power of nuclear fission and, even more worrisome, nuclear fusion, the elemental energy that fueled the sun, the fountainhead of all life on earth. What, then, would such men think of Ivry's audacious plan to obliterate the engine of this forbidden energy—and, God forbid!—maybe unleash it on the world? In a last, unintended irony that conjured exactly such prophetic warnings, the attack on Osirak was timed to commence exactly at sunset.

It was 1300 hours. Ivry and Eitan watched as the men filed into the briefing room, smiling and nodding hello to one another in the easy manner of a family gathering at breakfast time, despite an obvious tension in the air. General Ivry had personally picked each of the eight pilots for the mission. In a country that had already fought five wars in twenty-five years, the armed services were an elemental part of Israeli society. Most of the country's leaders and policymakers were former military men. Military service was compulsory for all Israelis at age eighteen—men served three to four years and one day a month in the reserves until age fifty-five, women two to two and one-half years and one day a month until age twenty-five. Young men and women shouldering M-16s were a common sight on downtown streets in Tel Aviv or Jerusalem. Everyone served. But those who chose to become IAF pilots, to become one out of ten who passed years of flying school and intense training, were an elite breed.

They wore with pride the berets and insignias of their command. They were recognized throughout the nation as belonging to an exclusive military caste. As they grew older, they would attend one another's sons' and daughters' graduations and weddings, and the births and bar mitzvahs of their children's children. Ivry and his team had spent a year together training in secrecy every day. They had shared jokes and meals. Had met one another's wives. Each man had become like a son to the two generals. And now they might never see some of their faces again.

The modeling experts in Operations who computed these kinds of things quietly projected at least two casualties—one due to equipment failure, one to enemy antiaircraft fire.

"I wish I were going with them," Eitan said, letting slip more emotion, Ivry thought, than he meant to.

The two generals moved to the front of the room and took their seats before the mission briefing, sitting just to the right of the podium and a huge map of the Middle East. One by one the operations specialists updated the pilots on the weather and flying conditions; General Saguy and military intelligence again covered the Saudi radar and AWACS patrols, the Iraqi airfields outside Baghdad, and the formidable antiaircraft and SAM (surface-to-air missile) battery emplacements surrounding al-Tuwaitha. The team leader rehashed the flight plan. They would navigate only a hundred feet above the ground over Aqaba, Saudi Arabia, and western Iraq. They were reminded to observe radio silence the entire journey. Each plane would carry only two air-to-air Sidewinder missiles instead of the usual four and no jamming devices to scramble MiG and SAM-6 radars. Too much weight. They had barely enough fuel to get to Baghdad and back even without the extra poundage. Each pilot had been requisitioned a day's ration of food and water, a pistol, five thousand Iraqi dinars, and PRCs, the electronic homing devices that would guide SAR (search-and-rescue) teams to their positions should they be shot down.

"But do not activate your PRC until nightfall," the team leader ordered. "We cannot take the chance your signal might be picked up by bad guys and the mission blown."

At 1440 hours the mission briefing ended and the pilots filed out the door. The sun was now high overhead, the desert air heavy. Inside the underground hangar the F-16s sat silently under the bright lights, lined in two rows, their noses down and brooding. Each pilot made a last visual inspection of his aircraft, then climbed the steel ladder up to the cockpit.

The crew chiefs followed the pilots up the ladders, carrying their flight helmets. With a farewell pat on the shoulder and wishes of "Good luck" from the chiefs, the pilots pulled down the glass-bubble canopies, the unique see-through feature that had given rise to the plane's nickname, dubbed by skeptical veterans, the "glass coffin." One at a time the F-100 Pratt & Whitney engines were lit. The high whine of turbines and sucked air created a deafening roar that shook the asphalt beneath. Inside the cockpits, the pilots went through the computerized BITS, built-in test systems, checking off the navigation, weapons, mechanical, and electrical systems before final takeoff. Then, slowly, finally, the fighters taxied up and out of the hangar and onto the tarmac at the head of the runway, staggered in two parallel lines, four planes to a line. The flight controller ran before the planes, wearing protective hearing mufflers and carrying red signal flashlights. The team leader gave the thumbs-up though the glass canopy. The mission was "Go."

The three generals stood off the runway, next to the taxi vans:

Ivry, the commander who had conceived the raid; Eitan, the general who had ordered it; and Saguy, the intelligence chief who had once opposed it. After years of planning and worry and failure, they were spectators now, impotently standing on the sidelines, each left to his thoughts.

Ivry squinted down the tarmac at the fighter-bombers, the wavy, superheated air from the jet exhausts obscuring the outline of the planes as though some flawed pane of glass had been dropped between them. Soon the Falcons would hurl down the runway and lift off to the sky, two at a time, climbing eastward, looking very much like the birds of prey they were named after.

Would their pilots come home safe? Would they be successful? What would the world think? What would be the final reckoning of this Raid on the Sun? Ivry thought.

The general turned wordlessly back in the direction of the command bunker, where he would wait and wonder what the night would bring.

CHAPTER 1

TERROR OF THE TIGRIS

Prepare whatever resources and troops
you must to terrorize the enemies of God.

—THE QUR'AN

Before the birth of its First Citizen, the flat, dusty village of Al Auja, just south of Tikrit and a hundred miles north of nowhere in the Mesopotamian desert, was best known to historians as the site where the vicious fourteenth-century Tartar chieftain Tamerlane chose to erect his infamous pyramid of skulls, a towering obelisk of death fashioned from the decapitated heads of thousands of slaughtered Persian soldiers. In an ironically unconscious homage, Saddam Hussein, who didn't know Tamerlane from Timbuktu, would one day commission his own public sculpture in Baghdad featuring two gigantic arms bursting through the sand brandishing a pair of crossed scimitars that crowned a similar pyramid of skulls fashioned from the helmets of thousands of slaughtered twentieth-century Persian soldiers, known now in modern times as Iranians.

It was into this savagely unforgiving desert that young Saddam, whose name in Arabic means to "strike" or "punch," was thrust on April 28, 1937, fatherless and penniless, to be reared in a mud-

and-straw house on the kiln-hot banks of the Tigris, without electricity, running water, or paved roads. Hussein would never forget his Tikriti roots. As though drawing inspiration from the land itself, he was mesmerized as a village boy by the country's ancient glory when it sat at the head of the Fertile Crescent, long before Abraham marched south from Ur in northern Mesopotamia to lay claim to the tribal homeland of the Semites. Much more than Iraq's later Islamic heritage, divided between the Sunni sects of the north and Shia of the south, Saddam identified with the country's pre-Arab Babylonian roots. He revered the great Babylonian king Nebuchadnezzar, whose golden age of prosperity had transformed ancient Baghdad into an intellectual center of trade and the arts, renowned throughout the Old World for such wonders as its legendary Hanging Gardens of Babylon. Even more impressive to young Saddam, Nebuchadnezzar was the last Middle Eastern ruler to conquer the Jews. Following a revolt in Palestine in 587 B.C., Nebuchadnezzar's army had destroyed Jerusalem, razing the First Temple and bringing an end to the kingdom of Judea. Thousands of Jews were marched in bondage back to Mesopotamia in what would become known in Talmudic history as the Babylonian captivity.

Hussein loved to recount the historic event to colleagues. And he would boast that someday he would follow in the footsteps of the legendary king to rule both the Middle East and Israel. Indeed, years later, after he had assumed regal-like powers, Hussein would embark on a Baghdad beautification program of public artworks, broad boulevards, and thousands of transplanted palm trees meant to evoke the great age of Nebuchadnezzar. Hussein named his sons Udai and Qusai, not names associated with Mohammed but with pre-Islamic Mesopotamia.

Maintaining fantastic dreams was especially important to young Saddam, who found his own reality nearly unbearable. Iraq was a tangled nation of tribes and clans and ethnic divisions. Where one was born and to what clan, or extended family, was extremely important, even in the poorer classes of Iraqi society. Surnames often derived from the village of one's clan. Most families could trace their forbears back for generations. Saddam's parents were not particularly distinguished members of either of their clans. Al Auja and Tikrit were considered backward, rural villages, and Tikritis of little account. Saddam's father's clan, the al-Majid, was considered lower caste.

Whether dead or just plain deadbeat—no one has ever seemed able to say for sure—Saddam's biological father left his son to be raised by his mother, Sabha, of the Tikriti Talfah clan, and a series of uncles, including one from his father's al-Majid clan named Ibrahim Hassan, who, regrettably, was better known to locals as "Hassan the Liar." (Tellingly, Saddam Hussein ultimately took his name from neither family—Talfah or Hassan.) Ibrahim eventually married Sabha and, with no trade and plenty of time, amused himself by beating young Saddam with a stick whenever he was bored, which, unfortunately for the boy, was often. He grew up mostly alone, forbidden to play with the other villagers, whom his uncle called "brigands," and who, in return, would taunt him for not having a "real" father. Saddam pined for the day he could escape Tikrit for a better world—a Mesopotamian world. In the meantime he bided his time, spending most days sitting by the side of the dirt road at the head of the village next to a fire pit with a red-hot poker, which he would stab into the stomachs of hapless village dogs that wandered by. This early cruel streak might have been occasion for worry, but on the sun-blasted streets of dirt-poor Tikrit it hardly went noticed.

Saddam finally caught a break when, in the fall of 1955, his mother's more prosperous brother, Khairallah Talfah, an ex–army officer turned hotheaded Arab nationalist and teacher, took Hussein along with his own son to Baghdad to attend secondary school. Saddam had just turned eighteen. Baghdad would change him forever.

In the early 1950s the city was a hotbed of ethnic and political radicalism. Iraq, which in Arabic means "the edge," was an amalgam of deeply divided tribes and ethnicities, the remnant of the defunct Ottoman Empire and Britain's Central Asian empire, which, following World War I, had been carved up into Iran and Iraq without taking into account traditionally and ethnically bound territories. Thus, most of southern Iraq, nearly 100 percent Shi'ite, had more in common with Iran than its Sunni "brothers" in the north. In fact, Iranian Shi'ites still revered two southern Iraqi cities as sacred religious shrines, including Najaf, the burial place of Mohammed's son-in-law, Ali (and the site of the horrendous terrorist bombing of its ancient mosque in August 2003). Meanwhile, distrustful of both the ruling Sunni and the southern Shi'ites were the northern Kurds, who were far closer in history and culture to the Kurdish tribes across the border in Turkey. By the 1950s Baghdad's tangle of ethnic divisions was further complicated by a slew of competing political parties, ranging from the Hashemite monarchists (the royal Arab family that ruled Jordan and whose scion, Prince Faisal, Britain had elected to rule Iraq in its stead), the right-wing Independence Party, and the centrist Liberal Party to the leftist People's Party, the Communist Party, and the secretive, socialist Arab nationalist Ba'th, or "Renaissance," Party.

Despite tutoring by his uncle, Saddam found it difficult to shed his peasant roots in the class-conscious big city, especially the crude accent that marked a rural Tikriti as unmistakably as a Cockney in St. James's Court. He failed to pass the entrance exam to join the prestigious Baghdad Military Academy. The stigma of outcast propelled Hussein, along with many of the city's disenchanted youth, toward the young, rebellious, socialist Ba'th Party.

Founded in Damascus by two Syrian intellectuals in the early 1940s, the organization espoused vaguely pan-Arab nationalist and socialist principles similar to Egyptian president Gamal Nasser's Arab Legion. But the party's immediate attraction to Baghdad's frustrated young rebels was its intense hatred of Western colonialism, especially what it saw as its expansionist guerrilla state— Israel.

Hussein hagiographies would later attribute his party association to his newfound belief in Islamic nationalism. In truth, the impressionable peasant's son was greatly influenced by his uncle Khairallah, who having been jailed by the British for his part in Baghdad's short-lived pro-Mussolini revolt in 1941, was the closest thing Saddam had to a hero. Khairallah mentored him in the tradecraft of Iraqi politicians: manipulation, intrigue, and anti-Semitism. Not one for mincing words, Khairallah's collective wisdom would later be published for the benefit of future Ba'thi in his book *Three Whom God Should Not Have Created: Persians, Jews, and Flies.*

Hanging around political meetings and outdoor rallies, Saddam eventually caught the attention of Ba'thist officials, who spotted a use for the young man's brooding, sadistic personality. They soon put him in charge of recruiting schoolmates and organizing local street bullies into brownshirtlike gangs to intimidate shopkeepers and the suburban middle class. Saddam became a familiar figure at Ba'th street demonstrations and public beatings, where he would stand off to the side, ordering his thug warriors into the fray or pointing out local tradesmen to be beaten. Things changed radically on July 14, 1958, when Gen. Abdul Karim Qassem and his "Free Officers" Brigade, backed by the Ba'th Party, marched into Baghdad and overthrew King Faisal and the fading Hashemite monarchy. It was a remarkably vicious coup, even by Iraqi standards, the carnage sinking to a new nadir when wild-eyed, rampaging mobs—waving sticks and swords and rushing up and down alleys looking for people to punish—discovered the whereabouts of the body of the just-murdered prime minister Nuri al-Said. The crowd promptly dug up the corpse and dragged it through the streets of Baghdad on a rope. (The scene would ironically be reenacted, at least symbolically, forty-five years later when, at the end of the Iraq War, Baghdad mobs, unable to find the real, live Saddam, made do by pulling down the famed statue of their erstwhile leader and then dragging it through the streets at the end of a rope.) Indeed, Saddam watched this day from the sidelines, but the rising twenty-one-year-old politico could not help but learn a valuable lesson he would not soon forget about the true nature of power in Iraq—and the loss of it.

One year later Saddam found himself waiting on a darkening Baghdad street not far from the West German embassy. He shifted his weight from one foot to the other, awkwardly trying to conceal the submachine gun beneath his cloak. How far Hussein had advanced in the Ba'th Party was confirmed by his selection as part of the hit team to assassinate the repressive Prime Minister Qassem, who had fallen out of favor with the radical Ba'this. As Qassem's car pulled to the curb on the other side of the street, a deafening blast of machine gun fire mowed down everything in sight, bodyguards and hit men alike, throwing the scene into bedlam. In the ensuing confusion the prime minister's bodyguards, recovering from their initial shock, shoved Qassem into a passing taxi, which rushed him safely to the hospital.

As the official legend would later explain the apparent screwup, Saddam, "when he found himself face to face with the dictator, was unable to restrain himself. He forgot all his instructions and opened fire." Saddam's job, in fact, was simply to provide cover for the hit men, who were to converge on the car from two sides and liquidate Qassem. Instead, Saddam panicked and opened fire on everyone, including fellow conspirator Abdul-Whahab al-Ghariri, who managed only two shots, one into Qassem's shoulder, before being dropped by Saddam's deadly fusillade. In the counterfire, Hussein was wounded in the leg and fled for his life, leaving behind al-Ghariri, who was quickly identified by Qassem's security team. Al-Ghariri's name led them to the rest of the team, and in short order Qassem's agents were hunting down one by one the surviving conspirators, including Saddam.

Again, according to the official Ba'thist story, Saddam, in pain and under sentence of death, dug the bullet out of his leg with a knife and, helped by kindly nomadic Bedouins and simple peasants, limped his way to Tikrit disguised as an Arab wanderer. In truth, a local doctor removed the bullet, and Saddam's uncle helped him cross the border by car into Syria and then finally to Cairo, where Ba'th Party loyalists took the young rebel in and enrolled him at the university.

For the next decade Hussein lived meagerly in a modest student tenement building not far from Cairo University, where he took up studying law. He was supported by a monthly government stipend granted by Egyptian president Gamal Nasser's regime, which had made Cairo a safe harbor and side station for revolutionaries and Arab nationalist dissidents from across the Middle East. For most of the next decade, Saddam lived cheaply and spent his time studying and working for the Ba'th Party. Occasionally, when the young Arab nationalist would run short of funds, the kindhearted *bawab,* or porter, of his building would loan him money. (Years later, after Saddam rose to power, he sent to Cairo for the old *bawab* and bestowed upon him a brand-new house as a reward for his earlier kindnesses.) The many dissident exiles in Cairo grew together to form a tight-knit community of idealists and revolutionaries. Sporting ethnic kaffiyehs and colorful regional dress from their various nations of exile, the foreign students, dissidents, and party chiefs gathered regularly in the capital's downtown cafés and coffeehouses to argue politics and promote socialism and pan-Arab nationalism. Among the most strident of the Arab one-worlders was Saddam Hussein, who lobbied loudly and tirelessly for his Ba'th Party back in Iraq.

In 1968, the Iraqi Ba'th Party finalized secret plans to take over the Iraqi government. Saddam promptly left his law studies and returned to Iraq to help foment the revolution. Hussein worked his way up the ranks to become head of the Ba'th security brigade, whose function was to eliminate enemies and protect party leaders, among them Saddam's cousin, Ahmed Hassan al-Bakr, a former schoolteacher who had become head of the Iraqi Ba'th. Adapting his own version of the carrot and stick—*tarhib* and *targhid* ("terror" and "enticement")—Hussein formed friendships and alliances when they furthered his career and broke them when they did not. He displayed a gift for understanding human nature at its most basic level and grasped early on that physical punishment was a great motivator, but *fear* of physical punishment was even greater. He found he had only to make one or two particularly dramatic examples and a reputation for his ruthlessness and retribution would grow by itself—the myth, in fact, far outstripping the reality. Years later, after he had become a complete tyrant, Hussein arrested a general, Omar al-Hazzaa, who had been heard to speak badly about him. He did not just sentence al-Hazzaa to death. Saddam first had al-Hazzaa's tongue cut out and then had his son executed as well. His home was bulldozed flat and his wife and surviving children turned out on the street. Such horrors, Hussein found, were a much greater deterrent than simply planting the general in the ground.

The macabre, almost routine cycle of bloodbaths in Baghdad

finally came to an end at three o'clock in the morning, July 17, 1968, when the Ba'thists stormed the presidential palace and ousted Qassem, and al-Bakr, the new prime minister and commander in chief, proclaimed the Age of Revolution.

It would be another year before Saddam would emerge from the shadows behind the al-Bakr throne. Then it was revealed that Saddam was second in command to his uncle, vice president of the secretive, all-powerful Revolutionary Command Council—and had been since the earliest days of the regime.

Halfway across the world, the carnage that roiled his homeland and filled his father's letters with worry seemed far away and almost dreamlike to Khidhir Hamza, as though he were hearing about someone else's Third World nation. The young middle-class Iraqi had lived in the United States for six years, studying at MIT and working for his doctorate in theoretical nuclear physics at Florida State University. Hamza's world had been filled, like all American students at the time, with campus antiwar demonstrations and the "flower power" culture. Indeed, Khidhir felt more American than Shi'ite Iraqi. To him the Ba'thi sounded like some organization from another century, if not from another planet. Hamza had received his doctorate and just begun teaching at a small black college, Fort Valley State in southern Georgia, when in 1970 he received a notice from the new Ba'thist Iraqi government. He was expected to return to Baghdad and repay his government student loan or his father would be held "accountable," a threat that in Iraq in those days could mean prison. It was a nightmare, an inconceivable turn of events Khidhir and his father had never even considered when they signed the loan document in 1962 under an entirely different regime. Only twenty-nine, unmarried, tall, with deep brown eyes and soft, light features, the young professor knew nothing about the Ba'th Party except what he had been told by friends—that "these guys are not fooling around. They kill people."

Hamza had no choice but to return to Iraq. He reluctantly resigned his teaching position, organized his affairs, shipped what few possessions he had to his father's home, and took the seventeen-hour flight to Baghdad, not knowing what to expect but fearing the worst. Bleary-eyed, Hamza stared out the porthole as his plane descended to the hot desert runway below. The Baghdad airport looked run-down, bleached of color and life by years of sun and wind and neglect. Carrying his own bags through the eerily deserted terminal, he caught a beat-up old Citroën taxi outside, which drove him to the Kuwait Hotel (the irony of the name would not be apparent for another twenty years, of course). There, a request awaited him from Dr. Ali Attia, director general of Iraq's impressive-sounding Nuclear Research Center of Atomic Energy, that he stop by for a visit.

Located south of Baghdad at al-Tuwaitha, past the al-Rasheed military base and the Iraqi army medical school and next to a tiny village-slum, the center was then a cluster of concrete government buildings surrounded by a high steel fence. Although al-Bakr was prime minister, everywhere Hamza looked on the center's walls hung portraits of Saddam Hussein, smiling under the thick Jerry Colonna mustache and wearing his trademark black fedora. Attia informed the nuclear scientist that he was to start as a researcher in the physics department, at $150 a month.

"That's only a tenth of what I was making in the United States," Hamza protested.

"We do not have the budget to pay more at present," Attia said. "But that will change."

Attia confided that Hamza was to be groomed to become his number-two man. Over the next weeks and months the scientist discovered he had joined an impressive team of Western-educated Iraqi scientists that included such respected researchers as Dr. Hussein al-Shahristani, Dr. Moyesser al-Mallah, and Dr. Abdullah Abul-Khail. All had been brought to al-Tuwaitha, he would discover, under the personal direction of Saddam Hussein. Though unbeknownst to Hamza at the time, Saddam, in fact, was not only vice president of the Ba'th Revolutionary Command Council, RCC, he had seen fit to appoint himself head of Iraq's atomic energy commission.

Al-Bakr was allowed to rule the far-flung Ba'th bureaucracy, but it was Saddam who plotted and wheeled and dealed discreetly behind the scenes for ten years to build Iraq into the dominant Arab power in the Persian Gulf—the modern Mesopotamia he had dreamed of. Far from the dumb thug his enemies liked to portray him as being, Saddam had impeccable instincts and a quick mind. In 1970 he foresaw a multipower world, with Iraq joining the Western powers as one of the pillars of global influence. Nasser had failed in his gambit to unite the Arab nations under Egypt. But Saddam would succeed where Nasser had failed, because he knew that the key was WMD, weapons of mass destruction—especially *nuclear* weapons.

Atomic weapons, in fact, became an obsession with Hussein. When a journalist once asked his son Udai what he wanted to be when he grew up, Udai had answered, "a nuclear scientist," eliciting an approving chuckle from his father. For Saddam, nuclear power was the ultimate symbol of the world player, a prerequisite for regional hegemony, and, of utmost importance, the "great equalizer" to finally match Israel's power.

In a little-known historical irony, the man who planted the seed of Hussein's fixation on atomic weapons was none other than Yasser Arafat, leader of the Palestine Liberation Organization. By 1970 the Palestinian intelligentsia had become the leading scholars of Baghdad. Palestinian refugees had flooded into Baghdad, along with most of the militia and their leadership, in the wake of Black September, when Jordan's King Hussein, thwarting a plot by the PLO to overthrow him, unleashed a bloodbath to drive the terrorist refugee camps from Jordan. Traditionally, Arab émigrés were desperate to disappear into their adoptive cultures, anxious to leave behind the bad memories of oppressive dictatorships. But the Palestinians, especially the educated classes, bound together as a minority, recruiting and fomenting the fight for the liberation of their homeland. This circle of energized scholars, educators, and propagandists became the hub of Baghdad's intellectual life, its café society.

One of the most talked-about books in the Arab universe in 1970 was *The Israeli Bomb,* written by a Palestinian-American academic named Fouad Jabir. Rumors had circulated for years that Israel had secretly produced an atomic bomb, but nothing had been proved. Not even United States intelligence knew for sure. Jabir's premise was that not only did Israel already have the "bomb," within ten years it would have a hundred atomic bombs. As long as Israel had this nuclear superiority, the Arab world would face a bleak future. Without a Muslim bomb and a "balance of terror," Jabir argued, Arabs would always be treated like second-class citizens, subservient to Zionists.

Eager to spread anything anti-Israeli, the PLO flooded the Middle East's urban centers with the book, and Arafat made sure the tome was brought to the attention of Hussein and the Ba'this through PLO operatives in Baghdad. For months Hamza saw stacks of the book piled in the Nuclear Research Center offices. *The Israeli Bomb* became a popular topic of debate throughout Iraq, where its message was tailor-made for Ba'thists like Saddam, whose dream was to lead the Arab nations to destroy Israel. Even more important, the book had provided Hussein with a blueprint on how to join the nuclear club. Ironically, as Israel would soon discover to its chagrin, he would do so by following, step by step, Israel's own model.

For Fouad Jabir, as it turned out, had been right all along— he had just been too conservative. Israel not only had an atomic bomb—it already had close to one hundred of them.

The year was 1956, and Israel's beloved "old man," David Ben-Gurion, had been unable to connect with President Dwight D. Eisenhower. The white-maned founder and leader of Israel for a quarter century embodied the passion of the freedom fighter with the soul of a rabbi, but his charms had eluded the U.S. president. Eisenhower had refused to form a security agreement with Israel. He had steadfastly adhered to a status quo policy in the Middle East, despite continuing signs of Arab aggression. He and the suite of Wall Street lawyers surrounding him seemed, if anything, to Ben-Gurion, more predisposed to the sheikhs who supplied the U.S. with oil, which drove its economy, than to Israel, which offered friendship and moral arguments.

The rebuff only deepened Ben-Gurion's continual sense of being alone. American journalist Seymour Hersh recalled a Ben-Gurion aide confiding to him once that the prime minister would sometimes cry out, "What is Israel? Only a small spot. One dot! How can it survive in this Arab world?"

As many an Israeli would note, to the West, Israel looked like David; to the Arabs, the nation looked like Goliath. But to the Israelis themselves, they felt more like Job. It was true that Israel had vanquished the combined armies of Jordan (then Trans-Jordan), Syria, Lebanon, Egypt, and even a token force from Iraq in its War for Independence in 1948, an event the Arabs call *al-Nakbah,* "the disaster." There was a sense of invincibility about the

Israel Defense Forces, whose soldiers and pilots were among the best trained in the world, but, ironically, there existed an almost equal sense of vulnerability. Like many of his generation, Ben-Gurion could not escape a feeling of doom, the conviction of *ein brera,* "no alternative," that his nation was surrounded by enemies who would never change and would never accept them, thus forcing the Israelis to do anything to protect themselves or face a second Holocaust, an Arab version.

Nasser had risen to the height of power in the Arab world by promising a day of reckoning with the "Zionist entity." Playing the Cold War chessboard, he had allied Egypt with the Soviet Union and received huge shipments of military aid, modern artillery, MiGs, tanks, and training, making his army the largest in the region. Egypt had been girding for war for four years, forming military alliances with the other Arab nations. Now, in early 1956, Egyptian forces had amassed on the Sinai border across from Israel. But still the United States and Europe refused to act. Finally, fed up with Washington and convinced that Israel could count on no one but itself, Ben-Gurion called on Foreign Minister Shimon Peres and Ernst David Bergmann, Israel's Oppenheimer, to fly secretly to Paris to request France's help in developing a nuclear reactor. The Israeli leader had concluded that the only way to ensure Israel's survival was by atomic bomb.

Six weeks after Peres left, in October, the war in the Suez broke out. Nasser continued to rattle sabers, moving infantry, tank companies, and Eygyptian MiGs far forward into the Suez Peninsula. Finally, he effectively declared war when he ordered a blockade of Israeli shipping in the Red Sea north of Sharm al-Sheikh at the southern tip of the Sinai Peninsula. Aligning with Israel, France agreed to sell Israel the nuclear reactor, and in a secret agreement joined the British in a pact to reoccupy the Suez Canal, which had been ceded to Egypt. The plan called for Israel to attack Egypt, and then, as a ruse to restore order between the warring Israelis and Egyptians, Britain and France would intervene and reoccupy the canal. Right on cue, Israel unleashed its tank corps and quickly cut a devastating swath all the way to the canal. France and Britain deployed to invade and capture the Suez, but the Soviets, smelling out the plan, threatened to dispatch troops to reinforce Nasser. The United States and overwhelming international pressure brought Israel to a halt. Israel and Egypt suspended hostilities, and a United Nations force was decamped to ensure the neutrality of the Sinai Peninsula. Britain and France were sent packing from the Middle East, as it would turn out, for good. The balance of power between the Middle East and the European continent was irrevocably altered. But France was still on the hook for a nuclear reactor, the state-of-the-art EL 102, which could produce 24 million watts of thermal power and, at full capacity, twenty-two kilograms of enriched uranium a year, enough to make four bombs the size of the one dropped on Hiroshima in 1945.

Groundbreaking on the reactor commenced quietly, without public notice, early in 1958 on a remote, restricted patch of barren desert in Dimona, near the ancient city of Beersheba in the heart of the Negev. It would soon be the most important piece of real estate in Israel. For the next ten years, hundreds of French technicians, Israeli scientists, construction workers, thousands of tons of equipment, and an unending caravan of covered trucks and earth-moving tractors moved in and out of the tiny watering hole. Dozens of U-2 overflights had aroused deep suspicions within the Pentagon and the NSA about what the Israelis were up to, but in truth, most of the people in Defense and State, as well as CIA, were sympathetic to Israel, even if it wasn't stated U.S. policy. Until there was confirmable proof, conventional wisdom said it was better to say nothing and wait. This cat and mouse game would continue for an entire decade.

By 1960, France's new president, Charles de Gaulle, began to have second thoughts about his nation's secret nuclear alliance with Israel. Effectively squeezed out of its onetime colonial playground in the Middle East, France was increasingly concerned about finding a cheap oil supply. De Gaulle, never overly warm to the Israeli state, leaned more and more toward the Arabs. He sent word to Ben-Gurion that Israel would have to reveal publicly that it had developed a nuclear reactor, or France would go ahead and divulge it. In December, Gaullists leaked the story of France's construction of the Dimona reactor to London's *Daily Express* anyway. Deeply hurt, Ben-Gurion was forced to reveal to the Knesset that Israel had constructed a nuclear research reactor in the Negev, but for purely peaceful means. To offset protests, Israel promised to allow inspection teams from the United States into Dimona to confirm that no weapons development was taking place.

A team of scientists and nuclear specialists from the Atomic Energy Commission (the precursor of the Nuclear Regulatory Commission) was duly sent to Israel. Climbing through the tunnels and excavations, the inspection-team members found no evidence that the reactor was being used to produce weapons. It was, just as the Israelis claimed, a twenty-four-megawatt research reactor. Everything was clean—in fact, incredibly clean, since some rooms showed evidence of fresh plaster and paint. What the American engineers and scientists did not know was that they were looking at a nuclear version of *The Truman Show*. On Ben-Gurion's orders, Israeli engineers had constructed a false control room, replete with fake control panels, a jimmied computer, and needles and dials displaying phony readouts from a putative twenty-four-megawatt reactor. The technicians had practiced the charade for weeks, making sure everyone knew their part. There could be no mistakes. The inspection team, after all, was not stupid.

But it *was* fooled. Completely. Had the inspectors ventured to check the reactor core, they would have discovered huge stores of heavy water, a tip-off that the reactor was being used far above its stated capacity and generating great quantities of potential plutonium. To avoid that happenstance, Israel claimed that the reactor was in full operation—far too dangerous to allow inspectors into the core.

In truth, in view of Israel's ultimate intentions, even the "real" reactor was something of a Trojan horse. Israel was not so much interested in the energy produced by the reactor, which was converted into heat and steam and, ultimately, electricity, but in the *by-product* of the energy's production—the spent uranium fuel from which could be extracted plutonium, the essential ingredient of an atomic bomb.

The physics involved in a nuclear reactor is actually fairly basic. A reactor consists of a containment vessel, bundles of fuel rods filled with pellets of uranium 235, control rods (typically cadmium or boron, which absorb neutrons), heavy water (deuterium oxide produced from normal water by a process involving electrolysis), loops of pipe to carry superheated water, and a steam turbine outside the reactor vessel to produce electricity. The primary energy comes from the fissioning, or nuclear chain reaction, caused by the uranium 235 atoms, which, in concentrated and contained form, emit neutrons traveling at the speed of light. These neutrons collide with other highly charged U235 atoms, splitting the atoms' centers, or nuclei, smashing them into fragments, and, at the same time, releasing heat and even more neutrons from the separated nuclei. The free neutrons collide with yet more atoms, creating a chain reaction that, if left unchecked, ultimately results in a nuclear explosion. But in a nuclear reactor, the fuel rods and pellets of uranium are immersed in heavy water, which absorbs neutrons and modulates the fissioning. In addition, control rods are inserted between the fuel rods to absorb even more neutrons, slowing down the rate of the splitting atoms or, when withdrawn, speeding the rate of fissioning. The heavy water and control rods allow nuclear techs to sustain what is essentially a controlled nuclear reaction. The immense heat produced by the continually fissioning uranium is transferred to the pipes of freshwater, which run through the reactor. Inside the pipes the water becomes steam, which is then used to drive the electric turbines outside the reactor.

What was important about a nuclear reactor in terms of building an atomic bomb was the by-product—the spent uranium fuel pellets inside the rods. While being bombarded with neutrons, the uranium becomes enriched with plutonium isotopes. This plutonium can be extracted from the spent uranium by a chemical process, then fabricated by a special machine into a metal form.

This was exactly what Israel was doing. Each weekly cycle in Dimona produced about nine "buttons" of pure plutonium, or 1.2 kilograms. It required roughly 11 kilograms of plutonium to produce one atomic bomb. Thus, approximately every ten weeks Israeli engineers had enough plutonium to create one more atomic bomb. For use in a bomb, the plutonium was shaped into a perfect sphere and surrounded by a high-explosive material. Triggered to explode inward in a precise sequence of nanoseconds, the blast would compress, or implode, the plutonium core into itself. The plutonium, like the U235 in the nuclear reactor, would begin discharging neutrons, but, unlike the controlled fissioning in the reactor, the neutrons in the bomb would discharge at an immensely faster rate—faster than they could escape from the core. Ultimately the pent-up energy would go "supercritical," bursting outward and producing the immense explosion and familiar mushroom cloud of the classic atomic blast.

A process of "boosting," that is, of inserting tritium extracted from heavy water into the warhead at the moment of fission, would flood the core with yet more neutrons and add an extra nuclear kick, dramatically increasing the bomb's explosive "yield." By the early fifties, physicists were already developing the so-called hydrogen bomb, a two-stage device that used the fission from an atomic bomb to compress and trigger the fusion of a second compartment of deuterium, a hydrogen isotope that burns as the primary fuel of the sun.

Israel's atomic bomb–making facilities were far too expansive to hide on the ground at Dimona, even with the fake control rooms. So the Israelis went underground. Near the Dimona reactor was a nondescript, two-story, windowless administration building sheltering an employee canteen, a shower room, an air filtration system, and a storage area. On the second floor was a secret bank of elevators, bricked over and hidden from view of the inspectors. The elevator shaft sank eighty feet beneath the floor to a secret, six-story underground laboratory known to the Israeli workers as "the Tunnel." A labyrinth of underground rooms and units, the Tunnel contained a chemical reprocessing plant and bomb factory where, beginning in 1965, the weapons-grade plutonium extracted from the Dimona reactor was fashioned into atomic warheads. Tritium extraction was done in Unit 92. Overlooking four floors where the plutonium was chemically extracted from the spent uranium rods, which cooled for weeks in water tanks, was a huge glass-enclosed control room, nicknamed Golda's Balcony in honor of the famed Israeli prime minister who frequently visited Dimona after taking office in 1969.

The bomb factory's existence would not be corroborated for another twenty years. But by the mid-sixties, reports of the reactor on the surface had made U.S. intelligence agencies suspicious. Early in 1965 the AEC and the CIA began rethinking the conventional wisdom concerning Dimona. After a decade of turning a blind eye on the somewhat troubling question of Israel's real intentions regarding the ultimate use of its nuclear reactor, the Defense Information Agency (DIA), AEC, and CIA began anxiously speculating on the primary source of Israel's U235 fuel—especially after two hundred pounds of enriched uranium shipped by Westinghouse Company and the U.S. Navy to a small Pennsylvania nuclear processing and fabrication firm called Nuclear Materials & Equipment Corporation turned up missing.

What flagged the attention of the AEC and CIA was the fact the firm's founder and director, Zalman Shapiro, the American son of a rabbi and Holocaust survivor, was a well-known, outspoken supporter of the Jewish state as well as an active member in the Zionist Organization of America. Even more intriguing, Shapiro counted among his closest friends Ernst Bergmann, the nation's leading nuclear scientist. After an investigation but without a lot of hard proof, the AEC charged that Shapiro's company, NUMEC, had diverted the missing uranium to Israel and then attempted to bury the missing inventory in its convoluted bookkeeping procedures. Shapiro vehemently denied all the charges, as did Israel. The AEC, FBI, CIA, even Congress conducted a panoply of audits, reviews, and criminal investigations for ten years. But in the end the case came down to supposition and some suspicious transactions. No hard evidence was ever uncovered that NUMEC had diverted anything to Israel, much less U235. The lack of proof, however, did little to save Shapiro's reputation. He lived out his life marked as a suspected agent for Israel.

The matter was soon officially forgotten. Whatever the truth, by the end of the decade, how Israel had attained enriched uranium was no longer of interest to anyone—except Saddam Hussein.

By 1971, Khidhir Hamza had been assigned to review and evaluate the history and operations of the Nuclear Research Center at al-

Tuwaitha and to produce a definitive Atomic Energy Progress Report. Before long, Hamza found himself involved in every aspect of Atomic Energy's business. What he discovered surprised him: for all Hussein's obsession with control, it was clear that Iraq had been taken for a ride by the superpowers. In the early sixties, Iraq's Atomic Energy (AE), under directions from Hussein, had purchased a small five-megawatt nuclear reactor from the Soviet Union. The sale was of little concern to the U.N.'s International Atomic Energy Agency (IAEA) because the reactor was too small to produce weapons-grade uranium, which could then be used to create an atomic bomb. In fact, the Soviets had refused to sell Hussein anything larger than the five-megawatt reactor. Ironically, the Russians, unlike many of the West's democracies, especially the aggressively competitive French, Germans, and Italians, turned out to be strict enforcers of the international nuclear nonproliferation treaties. However, seeing the perfect opportunity to make a good profit off of what they considered an unsophisticated and technologically impoverished Arab satellite state, the Soviets, in Hamza's estimation, had put together a package of mostly outdated nuclear and power-generating equipment, including, of all things, a boiler dating back to the 1930s. Bizarrely, instead of fixing an exact price tag on the reactor and for all of its various facilities and equipment, the Soviet nuclear agency charged Iraq by the *ton*. Accordingly, the Russians heaped as much equipment onto the deal as they could, padding the service contract with scores of redundant engineers, technicians, and untrained hacks who collected large paychecks for doing virtually nothing.

Padding the payroll and shipping ancillary machinery was easy, since it was never clear even to Iraqi administrators what exactly the Nuclear Research Center was supposed to be doing. Sitting atop the second largest oil deposit in the world, Iraq was hardly in need of nuclear power to run its electrical plants. In recent years, in fact, Atomic Energy had been used mostly to screen Saddam's dinner. Hussein demanded only the best of everything when it came to his personal comforts. The Iraqi leader had his food flown in fresh daily from Paris. On its sojourn to his state-of-the-art kitchens, the finest French beef, lamb, lobster, and shrimp were routed first to technicians at Atomic Energy, where the institute's multimillion-dollar X-ray and chemical-processing machines were used to check the victuals for poisons or dangerous metals that could harm the Great Uncle, as Saddam had taken to being called by loyal party minions. At the slightest doubt, the suspect delicacy would be sloughed off to local markets or restaurants. One well-known story recounted by Hamza recalled a day when Saddam came down with diarrhea. A squad of security guards stormed into the palace kitchen and held the cooks for hours kneeling on the floor with gun barrels pressed to their heads until a doctor examined the leader and declared it a common virus.

Hamza found that within a year of its purchase, the Russian reactor had begun to leak radioactive contaminated water. It seemed that when it came time to clean the reactor core, a normal process of maintenance, the Soviet technicians and sales personnel claimed that maintenance was not covered in the original contract. "This is your responsibility, not ours," the on-site engineers informed the Iraqis. The Iraqi technicians at Atomic Energy had no experience in maintaining nuclear reactors. Not one of its members had the vaguest idea of how to clean the inside of a nuclear reactor. Certainly there were no tools or provisions for such an undertaking at al-Tuwaitha.

Atomic Energy administrators elected to contract the work out to a private industrial cleaning service in Baghdad. The workers at the company were experienced in cleaning and scrubbing industrial warehouses and fabrication plants, even chemical laboratories and production facilities. But not one of them had ever seen a reactor before, much less cleaned one. In the end the maintenance workers relied on what they had always done. They entered the mathematically smooth, precisely engineered seamless environment of the reactor core and began scouring the pristine walls with wire brushes and industrial cleaning fluids, as though the reactor were just another dirty factory floor in need of a good scrubbing. Unbeknownst to the maintenance crew, or to the Iraqi engineers and nuclear techs who ran the reactor, the wire brushes scratched and grooved the pristine surfaces of the core's containment vessel. These minute divots and engraved lines created weak spots on the surface. When the reactor was activated again, the superheated steam and extreme temperatures soon corroded the breaches in the surface, eating away at the material until the core began leaking the moderating water around the fuel rods. When Atomic Energy complained to the Soviet techs that the core was leaking, the Russian liaison officers countered that they had had nothing to do with maintaining the reactor core and that the decision to hire an outside, incompetent firm to undertake such a delicate process was solely the responsibility of the Iraqis. Iraq was left with an undersized and now unusable reactor.

Khidhir Hamza put all the details of the ten-year Soviet administration and shepherding of Iraq's nuclear energy programs in his report. He handed the finished document to AE at the end of 1971. The reaction to the report was not long in coming. Saddam Hussein may have been a bit of a rube when it came to nuclear technology dealmaking, but he was a fast learner. Just months after receiving Hamza's paper, Hussein in early 1972 ordered all Soviet personnel out of the country. Simultaneously he froze the balance of the remaining payments due the Soviet Union and directed that it be held in escrow according to international procedure. He also informed Moscow that he would pay only five hundred thousand dollars—and he would pay that balance in Russia's own rapidly falling rubles. The deal was "take it or leave it."

His part in the humiliation of the Soviets made Hamza something of a local hero at Atomic Energy. The engineer was given a
raise and increased responsibilities. The Research Center's new director, Husham Sharif, a small, cultured man who had replaced
the lower-bred Attia, began currying favor with his new star scientist. Even Dr. Moyesser al-Mallah, the secretary general of Atomic
Energy, began dropping by Khidhir Hamza's small office for an occasional cup of tea. One evening, curiously, al-Mallah requested
that Hamza and Sharif come home with him after work so they
could talk in "private." Hamza drove to the man's house, located in
an upscale suburb of Baghdad reserved for officials of the Ba'th
Party. He felt anxious, wondering what al-Mallah could want that
was so important they needed to meet in secret. Hamza and Sharif
settled into al-Mallah's comfortable den and began the usual office
chitchat when Sharif suddenly changed the subject.

"What did you think of Jabir's book?" Sharif asked, referring to
the Palestinian's much talked about study of Israel's atomic bombs.
The question, apropos of nothing, alerted Hamza that this meeting
was a setup.

"I think it's ridiculous," Hamza replied. In truth, Hamza did not
believe that Israel had the capacity to produce enough plutonium
to make atomic bombs. And how could they have tested their de-
signs to make sure their bomb worked? Certainly Israel had no
Nevada test ranges. He looked at the disappointed faces of his
bosses. He had given the wrong answer. He was, he realized, being
arrogantly dismissive.

Al-Mallah, with some satisfaction, informed Hamza that, in
fact, Saddam believed every word of the book. Not only that, but
the Great Uncle had ordered the Nuclear Research Center to create an atomic bomb for Iraq. Al-Mallah then explained that if the
scientists at al-Tuwaitha could not show any progress, the Great
Uncle was liable to grow impatient, and that would be a danger to

all of them. On the other hand, if the the three of them could come up with a viable plan, then funds, resources, and prestige would flow to them all. But, for security reasons, they would have to keep this a secret between them. Hamza felt his chest tightening. Good God, he thought, how could they possibly build a nuclear bomb? What would the West do if they found out?

On the other hand, the idea of creating an atomic bomb from scratch took his breath away. It was truly Faustian: to be given every resource, the latest technology, the country's finest minds to compete against the West's best and brightest to build what was truly the ultimate prize in nuclear physics. And yet, what would he sacrifice—his morals, his professional ethics . . . his soul? Years later, Hamza would feel pangs of regret about his part in enabling Saddam's ambitious plans to become a nuclear state, but as a young scientist eager to prove himself, trapped inside Hussein's crazy world of intimidation and dreams of world power, he could not resist. Over al-Mallah's dining room table that night, Hamza, al-Mallah, and Sharif began planning the creation of the first Arab bomb. And, they agreed at once, they would follow the lead of the Israelis.

Once he had dispensed with the Soviets, Hussein began searching for a new partner and, like the Israelis before, he quickly discovered the French. For all their cultured sophistication, in truth, the French loved nothing more than a good bargain. And no one knew how to bargain better than Saddam Hussein.

Early on, Hussein had learned that people were motivated by two things: fear and greed, or at least the prospect of easy money. For the first, Hussein turned to his stick, the Mukhabarat, Iraq's sadistic secret police; for the latter, he relied on an oil-reserve carrot of $45 billion. He had the power and the wealth. What he didn't have was time. The ticking clock, as with all dictators, was his enemy—the one thing he could not control.

Indeed, Saddam's obsession with speed was a constant torture to Baghdad's construction industry. Neophytes to government service quickly discovered the dangers inherent in working for Hussein. Entifadh Qanbar—who years later would flee Iraq through the dangerous no-man's-land of the northern Kurdish border and ultimately return with the exile group headed by Ahman Chalabi in 2003—was a young, bright engineer working in Baghdad in the seventies. Short, dark, full of nervous energy, Qanbar looked like an Iraqi Joe Pesci. He had been hired to refurbish Baghdad's historical palaces as part of Hussein's vision to restore the city to its Mesopotamian glory. At the time, a friend from engineering school was bidding on his first government contract: a three-story government chemical plant in south Baghdad. Among many stipulations, the bidding specs for the plant called for a one-year construction schedule. Qanbar's friend had always been a bit of a character—roguish, a gambler who was not averse to cutting corners. In the army he would routinely forge weekend passes for himself at a time when the brutal officers of the Iraqi military were shooting soldiers for far lesser transgressions. Determined to win the bid, the engineer slashed his construction time to six months and submitted his estimate to Hussein Kamel, Saddam's son-in-law and the minister overseeing all military procurements. Following standard procedure, Kamel read the proposal, accepted it, and automatically cut the engineer's deadline in half, to three months, before sending it on to Hussein, who had to approve all government contracts. Hussein read the proposal, okayed it, then slashed the schedule in half once again, this time to forty-five days.

"You have forty-five days," Kamel informed the shocked engineer. "I'll give you all the help you need. You can change what you want, requisition whatever you need, charge whatever you decide. But you *have* to have this done in forty-five days!"

To his horror, the next morning a detail of Hussein's security men arrived and surrounded the engineer's work crew and the construction site. No one was allowed to leave until the job was finished. For the next two weeks the engineer and his crew—carpenters, masons, bricklayers, painters, laborers—worked day and night, eating and sleeping in shifts on the construction site. Dispensing with normal construction processes of erecting a building floor by floor, from foundation to roof, the crew poured the foundation, and while it dried, threw up all three floors at once—plus exterior brick walls and roofing, all braced by scaffolding and girders—and then allowed it to dry as one piece in place. Two weeks later they tore down all the scaffolding and retaining braces and there it was, a brand-new building.

The contract also called for a five-hundred-space parking lot. Normally such a lot would be graded out and then refilled gradually with dirt while being compacted every two feet by tamping machinery to ensure a stable foundation. Once solid and leveled, the asphalt would be laid and rolled flat. It's a time-consuming process. But with two days to complete the entire job, the engineer simply dug out a three-acre rectangular pit and then filled the hole in with thousands of cubic yards of concrete. Scores of cement trucks lined up for miles, pouring hundreds of thousands of dollars of concrete for twenty-four hours straight. When the concrete dried, it was asphalted over. Instant parking lot—and one of the most expensive pieces of land in Baghdad.

In fourteen days Saddam Hussein's military, biological, and chemical weapons division had a new research building. Qanbar's friend charged the government one dinar a brick—a rate that would translate in American dollars to charging three dollars apiece for twenty-five-cent bricks. Kamel and military procurement did not raise an eyebrow at the exorbitant fee.

With outsiders, Saddam's business strategy was less Draconian but just as direct: You give me what I want—hard-to-get items like tanks and uranium and nuclear reactors—and I will give you rich contracts—obscenely rich contracts. This was in essence what he told Jacques Chirac during the French prime minister's groundbreaking visit to Baghdad in early 1974. It turned out to be an offer the French P.M. could not refuse.

The infamous OPEC oil embargo of 1973–74 had just ended, sending gasoline prices to unimaginably high levels and shifting a trillion dollars of global wealth suddenly eastward. France was already dependent on Iraq for 20 percent of its oil. As part of the deal, Hussein offered France 70 million barrels of oil a year at present market prices for ten years. In addition, Iraq would purchase billions of dollars of French military hardware, including tanks, helicopters, antiaircraft missiles, radar, and one hundred Mirage F-1 fighters. Chirac practically trembled when Saddam threw in gratis contracts to purchase 100,000 Peugeots and Renaults in two blocks of 50,000 each. And as a final sweetener, the French would develop a planned billion-dollar lake resort outside Habbaniyah, the location of a large air force base west of Baghdad. In return, Saddam got his nuclear reactor.

In September 1975, Hussein entered Paris like a conquering pasha out of *1001 Arabian Nights*. Flanked by a troupe of barrel-chested bodyguards, he led a parade of festively clad Iraqi fishermen bearing flaming braziers of roasted Tigris River fish down the banks of the Seine. As news cameras rolled, Jacques Chirac and various government ministers of President Valéry Giscard d'Estaing's administration gathered around the Middle Eastern cooking demonstration, tittering and smiling gamely as they sampled bites of fish served on aluminum foil, Baghdad-style.

"*C'est bon,*" they declared, fastidiously wiping fish oil from their fingers with paper napkins.

The fishermen, their hair tousled and looking as though they had slept in their clothes on the plane ride over, moved self-consciously between the fish and the French, exchanging anxious glances lest someone make a mistake. Faux pas in the service of the Great Uncle could often be fatal. But nothing was amiss this beautiful fall night along the glittering bank of the Rive Gauche, while above it all Saddam looked on, beaming like the proud father.

Those of the educated class back in Baghdad would cringe in mortification at the television news images of their leader, like some cartoon Ahab, trying to impress the gourmand diplomats with fish in foil. But smiling in his trademark black fedora, Saddam was enjoying his own private joke. As wags later quipped, he knew Chirac and his entire cabinet would happily have eaten old tires from the Tigris if it would have bought them hundreds of millions of dollars in cheap oil.

Hussein's trip was the reciprocal visit to seal the deal struck in Baghdad. Hamza and his colleagues had picked out the perfect reactor for the Nuclear Research Center: the Osiris reactor, a huge, aluminum-domed, top-of-the-line research reactor, named for the Egyptian god of the underworld. France would oversee the production, shipping, and construction of the reactor and train Iraqi technicians in its operation. Ironically, as it turned out, many of the French companies contracted to do the work were the exact same government-approved outfits that had secretly built Israel's Dimona reactor a decade earlier. France also expanded the original nuclear trade treaty to include yet another, smaller research reactor, "Isis," named after Osiris's wife, which would be erected alongside Osiris. Finally, in a rare and controversial decision, France agreed to supply Iraq with seventy-two kilograms of highly regulated enriched, or "weapons-grade," uranium for start-up fuel. This last agreement quickly caught the attention of the U.N.'s International Atomic Energy Agency, which kept a keen watch on any movement of U235 because it could be readily converted to use in an atomic bomb.

The reactor "listed" for $150 million. The price tag for Saddam was $300 million.

"We were happy to pay," Hamza would recall later. "After all, who else was going to sell us a nuclear reactor?"

Euphoric, Hussein rechristened the nuclear reactor *Osiraq* (incorporating the name *Iraq*), or "Osirak" in English, and the Nuclear Research Center "Tammuz," after the Arabic word for *July*, in honor of the month of the Ba'th revolution. Tammuz would form the centerpiece of Iraq's new nuclear energy industry centered at al-Tuwaitha, "the truncheon," in the brown flatlands of the Tigris.

The two Israeli generals, David Ivry and Raful Eitan, stared in silence at the row of grainy eight-by-tens, dealt like a poker hand on the table before them—aces and eights, a dead man's hand. Smuggled out of Iraq at great personal risk by Mossad agents, the photographs showed a veritable Nuclear Oz populated by steel-and-glass laboratories, a nuclear fuel reprocessing unit, modern administration buildings, a square mile of electrified fences, and, rising Venuslike in the center of it all, the huge, gleaming aluminum dome of the Osirak nuclear reactor.

Taken from ground level at al-Tuwaitha, the blowups were incontrovertible proof that Saddam Hussein's blueprint for an ambitious, modern nuclear program was proceeding at an alarming pace. Israel had known about the center, of course: Mossad had alerted Yitzhak Rabin to the possibility back at the time French prime minister Jacques Chirac first visited Baghdad in 1974 to discuss the trade treaty between France and Iraq. At the time, Israeli prime minister Rabin had called for Jewish-American organizations to pressure the Ford administration to help kill the deal. Defense Minister Shimon Peres had personally appealed to his close friend Chirac to cancel the contract. But the French could not bring themselves to abandon such a fat cash cow. Chirac reassured Peres that perhaps he could do "something" later, after the French national elections. In the end, Rabin decided to "wait and see."

Now, three years later, in May 1977, it was clear Hussein had much bigger plans than a simple research reactor. Israeli intelligence estimated Osirak would go "hot," that is, be fueled with radioactive uranium, within three years, four tops. Israeli scientists figured the reactor would produce enough enriched weapons-grade uranium to build two or three Hiroshima-size bombs a year. The contingency people calculated that one "small" atomic bomb dropped on Tel Aviv would kill at least one hundred thousand people.

Begin had just defeated Israel's liberal Labor Party to become the conservative Likud Party's first prime minister, and he quickly made dealing with al-Tuwaitha one of his government's first pieces of business. Thus, this secret Sunday morning meeting at the prime minister's heavily guarded offices in Jerusalem. Seated before him, along with Eitan and Ivry, was Begin's new "shadow security cabinet": Minister of Defense Ezer Weizman, large and bilious, one of Israel's founding fathers; Deputy Prime Minister Yigael Yadin; Military Intelligence chief Yehoshua Saguy, heavy eyebrows and brush mustache framed by a round face with perennial raccoon circles under his eyes; Foreign Minister Yitzhak Shamir, a no-nonsense military general; Agriculture Minister and legendary slash-and-burn tank commander Ariel "Arik" Sharon; and, finally, the chief of Mossad (officially the Institute for Intelligence and Special Operations), Yitzhak Hofi, tough, compact, and stubborn as the craggy Jerusalem pine.

It was obvious to everyone present that diplomacy had failed badly with Hussein. The United States and Britain had expressed official diplomatic "concern" about the sale of a nuclear reactor to

Iraq, but the U.S. was not keen on a showdown with the country. Hussein had begun to distance himself from the Soviet Union and encourage trade with the West. Iraq was importing more domestic goods from America than from the Soviets. Already, trade had reached some $200 million. Within two years, that figure would triple and, it was estimated, there would be two hundred American businessmen stationed in Baghdad. Having been blackmailed with an oil embargo, Europe was in no hurry to provoke the Arabs again. Certainly, France, which was making billions of francs on its nuclear trade with Iraq, had no intention of stopping work.

Israel would have to deal with Iraq alone. But what were its options? Iraq was one of the richest nations in the Persian Gulf, with a GNP of $18 billion—ten times the size of Israel. It had powerful allies, including the Soviet Union and the Arab Rejectionist Front, an organization of Arab nations, including Syria and Yemen, dedicated to the destruction of Israel. Iraq's army boasted 190,000 men, 12 divisions, 2,200 tanks, and 450 attack planes.

Surprisingly, the two intelligence chiefs, the IDF's General Saguy and Mossad's Hofi, as well as Begin's own Deputy Prime Minister Yadin, vehemently opposed any type of military raid. Such violation of a nation's sovereignty was tantamount to an act of war, they argued. It was too risky and there were too many unknowns. And besides, who knew for sure whether Iraq was truly capable of building an atomic bomb? It required a sophisticated technological and educated infrastructure, which Iraq clearly did not possess. Eitan and Ivry, joined by Sharon, countered that Israel could not afford to wait and find out.

Hofi's stubborn, mulish eyes clamped on Rabin.

"You run a much greater danger of alienating America than of destroying Iraq's reactor," he announced.

"What help will they be if he creates an atomic bomb?" Ivry countered.

Voices grew louder around the room. Though nothing compared to the Israeli Knesset, where parliamentarians routinely screamed at one another at the top of their lungs, hurling insults and threats, the meeting was nonetheless becoming tense and uncomfortable. These men had known and fought beside one another, literally in the trenches, for decades. But the critical nature of the "Arab nuclear question" and how to deal with it had profound and imponderable ramifications—and it cut to the bone of national survival.

The present government, it became immediately clear, was dangerously divided over how to handle Iraq's nuclear threat. Indeed, Hofi found himself ruling a house divided at Mossad. Most of his department heads supported a raid. Since when, they observed, did Israel care what Europe thought of its policies? Where were their "friends" in '67 and '73? David Biran, the head of Tsomet, Mossad's recruiting department, was already moving ahead with preparations for some kind of intervention by force at Osirak and had ordered the Paris station to find an Iraqi candidate working at France's Sarcelles nuclear facility, which was overseeing the construction of the reactor, whom they could recruit . . . or compromise.

Shocking the cabinet, the usually hawkish, shoot-from-the-hip prime minister, his shiny bald head and steely black eyes flashing around the table, announced he would not approve *any* military operation unless he had 100 percent backing of the entire cabinet. Rabin's election in 1974 had been partially the result of the continuing ideological temblors shaking Israeli politics ever since the trauma of the '73 Yom Kippur War, when Israel, after fatally misreading Arab troop movements along its borders, found itself in danger of being overrun by Syrian forces during the first three days of fighting. Begin, the hard-line general and infamous Irgun head, had been elected to ensure that such a disaster never happened again. But the kind of raid the cabinet was now contemplating, the

The thunderous roar of eight Pratt & Whitney jet engines firing up inside the cavernous underground hangar vibrated all the way through the crew chief's safety earmuffs, seeming to make the foam-rubber-lined earphones jump right off his skull. The F-16 maintenance chief swore he could actually see the sound waves rolling out of the exhaust burners. He glanced up at the cockpit of No. 106. Through the glass canopy the pilot gave him a thumbs-up. He was ready to taxi up the ramp to the runway for takeoff position. Where the pilots were going, the warrant officer had no idea. This was a "black operation," conducted in complete secrecy. The entire base had been locked down like a prison since Friday. All flight and support crews had been warned by security officers not to talk to anyone and to ask no questions.

Now it was T-minus 1:05 and counting. Takeoff was 1600. The maintenance chief could feel a knot in his stomach. Something big was happening. He ducked under the wing on the three o'clock side for final preflight inspection. He had already rechecked the plane's parachute fittings and affirmed that all the safety pins were pulled so the ejection seats were armed and ready. The rocket beneath each seat would shoot the pilots several hundred feet into the air above the aircraft in case of bailout. All the pilot had to do was pull the ejection handle. Now, the crew chief looked for hydraulic leaks, fuel leaks, damage to the fuselage. He checked the tire pressure and that the two-thousand-pound MK-84 gravity bomb was secured in its release clip and that the safety pin was still in place. He scanned the external fuel tank, hung between the bomb and the fuselage. The tank held an extra 370 gallons of fuel. He had not seen many external fuel pods. Nearly all IAF combat missions and patrols, even during hostilities, occurred within or just across Israel's borders. Such a long-distance strike was rare—maybe a first. The warrant officer and the other crewmen could not help but guess where the pilots were going. Most thought deep inside Syria—or maybe Libya. He thought it would be east.

He did a final visual of the Sidewinder missile mounted on the wingtip, looking for loose clips or unattached electrical wires that could cause the air-to-air heatseekers to malfunction—and perhaps lead to the death of the pilot. He repeated his inspection under the opposite wing, then gave an "all clear" sign to the pilot before ducking under the landing gear to pull out both chocks wedged beneath the tires. Unfettered, and with another deafening whine from its engine, the 106 Fighting Falcon inched forward, gradually gaining speed as it climbed the ramp out of its hidden nest to the tarmac outside. The crew chief walked beside the plane, blinking back the blinding rays of the summer sun, staring at the eastern sky.

CHAPTER 2

MISSION IMPOSSIBLE

First say to yourself what you would be,
then do what you have to do.

— EPICTETUS

The drivers of the two cargo trucks bounding across the French countryside carrying the Mirage jet engines from the Dassault-Breguet plant for delivery to a warehouse in the tiny Mediterranean port town of La Seyne-sur-Mer barely noticed when a third nearly identical container truck pulled onto the highway behind them and joined their caravan along the route to the French Riviera. The convoy stopped outside the main gate of a heavily fenced compound while the guardhouse security officer checked the paperwork of the first driver.

It was April 6, 1979, and three guards were on duty this evening, one of them a new employee with impeccable credentials who had been hired just days before. They all worked for a private French security company contracted to protect the compound. The guards, more concerned with shipments going out than coming in, waved the trucks through, including the third truck, which was ferrying a large metal shipping container. This truck turned off from the other two and pulled up outside a huge storage bay. The bay gate had been unlocked earlier by the new guard. Climbing down from the cab, the driver momentarily surveyed the darkened compound, then moved to the back of the truck and unlatched the doors to the metal container. Six men, all dressed in street clothes, quickly dropped to the ground. Five of the men were *neviots* (Mossad break-in specialists trained in sabotage and bugging), the sixth an Israeli nuclear engineer.

Inside the bay, crated and marked for shipment to Iraq, were the finished cores to the Osirak nuclear reactor, arguably the most critical components of the reactor. Crated nearby were more reactor parts and a huge metal block designed to house atomic batteries. Alongside the Iraqi shipment was a device for loading nuclear fuel into a reactor, scheduled for shipment to a Belgian company, and a specially designed lid to a container to store radioactive materials, ordered by a West German firm.

As the Israeli nuclear engineer quickly pointed out the most damaging places in the cores to plant five plastic explosive charges, outside the compound gates a crowd had begun to form. Across the street from the guards, a strikingly attractive woman had been suddenly brushed by a dark late-model car as she crossed the street. Whether injured or not, she was decidedly not dumbstruck and immediately began shouting obscenities in French at the shaken driver, drawing the attention of passersby and the security guards, who left the gate and jogged toward the woman to see if she was hurt. As the guards crossed the street, a deafening explosion behind them shook the village like an earthquake, blowing out windows blocks away from the plant and engulfing the shipping warehouse in flames. By the time the gendarmes and fire trucks had raced to the compound, both the car and the woman had disappeared into the night.

The twisted nuclear-fuel loader destined for Belgium and the

West German container lid were unsalvageable. Both of the Iraq reactor cores showed hairline fractures. Designed to withstand intense heat and radiation, the cores had been manufactured to exacting specifications. The slightest fissure could lead to a meltdown. French investigators estimated the damage at $23 million, U.S.

The French were curiously unapologetic. Dr. Khidhir Hamza and the Nuclear Research Center were informed that the cores would take two years to replace. They could be put online, but they would crack eventually. If Iraq wanted them "as is," it would have to sign a waiver releasing the French from all responsibility. Ultimately, Iraq's atomic energy officials, knowing the Great Uncle's obsession with deadlines, decided to accept the cores the way they were, cracks and all. The French agreed to perform what repairs they could.

French officials were closemouthed about the incident, and a police blackout was imposed on the media. Meanwhile, the French began a "below the shadow line" investigation. Immediately suspect were Libya, the PLO, the Russians, the Israeli Mossad, and even their own French secret service, which had been known to have grave misgivings about the Paris-Baghdad treaty from the beginning. After the blast, an anonymous caller from an environmental organization identified as Le Groupe des Ecologistes Français, a group no one had ever heard of before—and would never hear of again—telephoned the French daily Le Monde, claiming it had bombed the plant at La Seyne-sur-Mer "to neutralize machines that threaten the future of human life." A week earlier the near-meltdown at the Three Mile Island nuclear power plant in Harrisburg, Pennsylvania, had sparked antinuke demonstrations around the world. The caller referred to this accident, saying the group had turned to action "to safeguard the French people and the human race from such nuclear horrors."

French gendarmes dismissed the claim. For one thing, the explosives work had been too professional. And there was no history of the group. But the police had little else to point to, as there were few witnesses, no suspects, and vague motives. With few facts and French law enforcement refusing to comment further, the Paris media were left to speculate wildly about what had happened. *France Soir* maintained that "extreme leftists" had carried out the sabotage. *Le Matin* chose the Palestinians; the weekly *Le Point* laid it at the feet of the FBI. And, of course, everyone considered Israel and the Mossad. But where was the proof?

Back in Tel Aviv, Yitzhak Hofi smiled as he read the various media accounts, especially *Le Monde*'s exclusive report on the genesis of the mysterious, militant ecoterrorists, Le Groupe des Ecologistes Français. According to ex-Mossad officer Victor Ostrovsky, Hofi was especially fond of that article: after all, it was he who had personally made up the name of the "terrorist" group.

The sabotage at La Seyne-sur-Mer yielded an unexpected dividend for Israel. Media attention once again had focused the spotlight on France's controversial nuclear partnership with Iraq. Why was France helping an arguably rogue nation like Iraq achieve nuclear capability? And what did Iraq, floating on a sea of oil, need with a nuclear reactor? To blunt the growing international grumbling, Chirac announced that France would supply Iraq with only a low-grade uranium not suitable for weapons use, so-called caramelized uranium. Instead of the enriched weapons-grade U235 uranium promised in the original deal, the caramelized uranium was less than 40 percent pure and, though radioactive enough to power a reactor, it was unsuitable for the production of plutonium.

In Iraq, the bombing sent a chill through the scientists working at the Nuclear Research Center. Unlike the French, Khidhir Hamza and his colleagues had no illusions about the identity of the saboteurs: everyone suspected immediately the hand of the

Israelis. The notion that Mossad's deadly eyes had been turned on them caused a great deal of anxiety. Who could say whether the scientists themselves would be the next target? In the meantime, work went on as before. The Seyne-sur-Mer explosion delayed the installation of the cores for several months, an annoyance to be sure. But it had failed to destroy them, and construction at Osirak remained more or less on schedule.

Hofi and Mossad still had work to do.

In July 1979, just months after the explosion at La Seyne-sur-Mer, Iraqi president al-Bakr suddenly—and surprisingly—announced his retirement. Saddam Hussein immediately accepted the presidency. He was supreme leader, president of the Revolutionary Council and the Ba'th Party, and head of the army for life. The new title seemed to fill Hussein with a renewed viciousness.

Soon after followed the infamous Night of the Long Knives. The story had started as only a rumor whispered among Baghdad's party faithful until a videotape of the unbelievable event surfaced and circulated among the upper classes. Using the pretense of an attempted Shi'ite assassination of his longtime deputy Tariq Aziz, Saddam had ordered a special assembly, calling together hundreds of deputies, ministers, and members of his ruling Ba'th council. At the grand convocation, Hussein took the podium and announced that the government had been betrayed. As he spoke, security guards and agents of the Ba'th Party's dreaded secret police, the Mukhabarat, moved to seal off all the doors in the room. Then, one by one, sixty deputies and ministers, mostly Shi'ite, were called by name to the podium to confess to the room their treason and then, by way of apology, to recite the Ba'th Party oath: "One Arab nation with a holy message. Unity, freedom, and socialism!" When the oath was finished, the bureaucrat, pale and shaking, was led out a side door to a patio, where he was shot to death on the spot. Soon bodies were piled high on the bloody terrazzo. To prove their loyalty, factotums and party hacks—some grandfathers in their sixties, shaking and physically ill—were forced to pull the trigger on their former friends. The scene was straight out of some Brueghelian vision of hell. On the videotape, which Saddam personally ordered to be recorded, the Great Uncle could be seen laughing as the frightened men were marched away to death or prison.

Cut off from normal people, sleeping in a different palace or bunker every night, always fearful of revenge by a survivor or a child of a murdered adversary, Hussein grew more paranoid and eccentric. To confuse enemies he used doubles, men who had undergone plastic surgery to look like him. On the rare occasion he went out to dinner—even at the exclusive private Hunting Club in Baghdad—Hussein's security men would first storm the kitchen and then observe every step of the cooking process, checking for poisons. Obsessed with germs, like Howard Hughes and Hitler before him—Saddam allowed no one to touch him. If a caller forgot himself and tried to shake the Great Uncle's hand, bodyguards would billy-club him to the ground before the outstretched hand could violate the Great Uncle. A guard stood duty outside his offices with a doctor's penlight, checking noses and throats to ensure no one with a cold or the flu passed by.

Stories like these were making the scientists and engineers at Osirak increasingly paranoid themselves. But Saddam more or less left the nuclear scientists to their work. Then one day, in early December 1979, a caravan of black Mercedeses came racing down the road from the main gate at al-Tuwaitha and pulled up to the curb in front of Atomic Energy's administrative offices. Men in black suits and armed with submachine guns emptied from the cars and quickly sealed off the building. German shepherd police dogs were led through the hallways, sniffing, straining at their leashes. It was obvious: the Great Uncle had come to visit.

Deputy director Abdul-Razzaq al-Hashimi watched nervously as security agents entered his offices and ordered him to round up his top scientists. He quickly obliged. The room soon filled with nuclear engineers, physicists, and directors, including eminent scientists Dr. Hussein al-Shahristani, Dr. Jaffar Dhia Jaffar, and Humam al-Ghafour. Khidhir Hamza had flown to New York days earlier to attend a United Nations nuclear energy conference, but he would hear the harrowing details when he returned.

Saddam Hussein, histrionically, strode into the room without preliminaries. Guards pulled the doors shut behind him.

"When are you going to deliver the plutonium?" he asked the assembled scientists straight out.

An awkward silence hung in the room.

"I *said*," he repeated, "when are you going to deliver the plutonium?"

"Plutonium . . . for what?" AE's director, al-Shahristani, finally replied.

Saddam looked at him, annoyed. "When will you deliver the plutonium for the *bomb*?"

"Bomb? We can't make a bomb . . ." al-Shahristani almost stuttered. "Well, theoretically, we could, I suppose, if we had enough plutonium . . . but there are nuclear nonproliferation treaties . . ."

"Treaties are a matter for *us* to deal with," Saddam cut him off. "You, as a scientist, should not be troubled by these things. You should be doing your job and not have these kinds of excuses."

Hussein stared at the group of scientists, who all stared at the floor. Finally, he seemed to make up his mind about something, then turned and walked out the door.

The following day al-Shahristani was not at work. He was not seen again in Tammuz. In fact, as Hamza would learn later, he was jailed, first in Mukhabarat headquarters in Baghdad and then in Abu Ghraib prison outside the capital. Two days later Jaffar Jaffar was also picked up and jailed. When he returned from New York, Hamza was put temporarily in charge of the nuclear reactor program. Hamza had been let in on Hussein's ultimate plans for the Nuclear Research Center years earlier in the front room of al-Mallah's home, but by December 1979, few scientists working at Atomic Energy had any illusions about the real purpose of their work.

New equipment continued to arrive weekly. The Rome-based nuclear manufacturing firm SNIA Technit, following France's lead, had sold Iraq a critical chemical reprocessing unit used to extract weapons-grade plutonium from spent uranium fuel rods. Iraq was meeting with West Germany and Brazil about importing uranium ore and purchasing more nuclear reactors. A report by AMAN, the intelligence branch of the IDF, stated that one prospective deal between Iraq and Brazil called for the South American country to build *nineteen* nuclear reactors for Saddam.

Butrus Eben Halim was an unremarkable, henpecked, forty-two-year-old professional with no children and predictable habits. Every morning at the same time, at the same stop, he caught the same bus from Villejuif south of Paris to the train station at Gare Saint-Lazare Metro. The most interesting thing about him was that he was an Iraqi scientist working at the French nuclear reactor at Sarcelles. It was no surprise then that Halim was immediately intrigued by a rakish Englishman named Jack Donovan, who raced around Paris in a red Ferrari with an ever-present blonde in the passenger seat. Halim noticed him driving by the Villejuif bus stop on numerous occasions. So it was natural that one day, when the Englishman pulled up to the curb, asking if Halim had seen a blond woman waiting at the bus stop, the Iraqi would quickly fall to his charms. The two men struck up a conversation, and Donovan offered Halim a ride to the train station. By the time he had dropped Halim at the Gare Saint-Lazare station, Donovan had begun a friendship with the impressionable Iraqi.

It was exactly what Mossad's Paris station head, David Arbel, had counted on. Not long after Arbel, a distinguished, urbane man with white hair and impeccable manners, received the Tsomet request to find an Iraqi recruit, a *sayanim* (sympathetic Jewish volunteer) working in personnel at Sarcelles provided Mossad with a photocopied list of the names of all Iraqi scientists working at the plant. The personnel list had been double-coded at the Mossad station, located in the heavily reinforced basement of Israel's Paris embassy, using a system that ascribes each phonetic sound of a word a corresponding number. For instance, the sound "ah" in the word *about* might be a "2," the sound "bout" a "3." Thus, *about* would be 23. The message was then sent encrypted to the research departments at Mossad and AMAN. None of the nondescript scientists at Sarcelles registered a hit on the intelligence computers, so Halim was targeted in the spring of 1978 as a "hit of convenience," and a *yarid* (team of break-in, bugging, and security specialists) was assigned to work him.

A field officer, or *katsa,* observed him for a week, at first using "motionless following," that is, watching Halim in stages rather than tagging him. Meanwhile, a *Shicklut* team broke in and bugged his apartment with listening devices in order to learn about his personal life and piece together a profile. And, of course, to ensure that he was not under surveillance by another organization. The *yarid* would need as much information as possible, for this operation had to be a "cold approach," and recruiting foreign nationals was delicate work.

According to Victor Ostrovsky, the approach was made in

August by an experienced field agent he identified as "Ran S." It was Ran, Ostrovsky said, who posed as the rich and successful entrepreneur "Donovan," involved in business from London to Libya. A Mossad phone tap had revealed that Halim's wife Samira was returning to Iraq to visit relatives for the fall. Donovan took advantage of Halim's bachelor status to invite him to dinner and walks along the Champs-Elysées. He took him to fine bistros and clubs and to his luxurious hotel suite at the Sofitel Bourbon. They smoked cigars and drank fine wine, something that, as a Muslim, Halim was not used to. But under his new friend's influence, Halim loosened up and began to enjoy himself. One night after some heavy drinking, Donovan fixed the two of them up with some girls, then, making a phony excuse, left Halim behind with a young French prostitute, Marie-Claude Magalle, who was frequently employed by the Mossad for such jobs, though she had no idea that the organization hiring her was in reality the Israeli secret service.

Finally one night, Donovan invited Halim along on his latest business deal: a scheme to supposedly buy old shipping containers and then resell them to African nations to be used as temporary housing and storage. As Donovan feigned to close the transaction, Halim, aided by some obvious staging by the *katsas*, noticed that the bottoms of some of the containers were badly rusted. He pointed the damage out to Donovan, who then finagled a discount out of the sellers. When the deal was completed, Donovan celebrated by giving his shocked but grateful new friend a thousand dollars for his help. Though Halim, of course, could not know it, Ran was gradually binding the Iraqi scientist to him using the three time-tested hooks of the spy trade: sex, money, and emotional motivation—in Halim's case, excitement and friendship. He was now ready for the trap, for some real spy business, some *tachless*.

One night, some five months after they had met, Halim was

having dinner with Donovan and noticed his English friend seemed down. Donovan explained that he was having trouble with a huge deal: he had contracted with a German company to sell pneumatic tubes for shipping radioactive medical materials. The tubes were supposed to have been inspected by an English scientist, but he had disappeared. And now it looked like the deal would, too. Halim, taking the bait, spoke up.

"I can do it. I am a nuclear scientist."

Surprised, Donovan gratefully accepted his help. The next morning he flew them both to Amsterdam to meet the German businessmen, a Mossad case officer and an Israeli nuclear engineer, according to Ostrovsky, posing as "Mr. Itsik" and "Mr. Goldstein." After successfully concluding their business, all four went out to dinner. Saying he had to make a business call, Donovan excused himself from the table. Goldstein and Itsik took the opportunity to begin casually querying Halim about his relationship with Donovan and about his work. When Halim divulged that he was working on the Iraqi nuclear project, the two businessmen were taken aback. In an incredible coincidence, they told him, they were at present working on a deal to sell nuclear power plants to Third World nations.

"Your project would make a perfect model for us to use to sell these people," Itsik remarked. "We could all make a fortune."

He leaned closer across the table to Halim. "But we have to keep this between us. Donovan will just want a piece of the action."

Halim was reluctant to supply any plant materials at first, but the two agents worked on him. In the end they convinced him that he had nothing to lose. After all, they only wanted a model of the plant. Nuclear reactors were not exactly state secrets, were they? Donovan paid Halim eight thousand dollars for his services and returned to England. As agreed, Halim went back to work at

Sarcelles and provided his new partners with a layout of the nuclear plant at al-Tuwaitha.

The plant outline of al-Tuwaitha showed the schematic of Osirak, the chemical reprocessing plant, the smaller Isis reactor, the administration buildings, and an underground tunnel leading off the main reactor used to channel off free neutrons for further experiments. Paris head Arbel sent the plans back to Hofi in Tel Aviv by armed carrier. There the plans were pored over by IDF intelligence and the IAF, including Ivry. But AMAN and the IDF needed more information. Ran, it was decided, would stay out of the picture while Halim's new handlers, the Germans, would use him as a "lead" to recruit another more senior scientist or administrator.

Halim was paid generously, but his new friends were much pushier than Donovan. They began demanding more details about the building of the reactor: its capacity, a timetable of shipments. Where were the parts stored before shipping? What was the date Osirak would go "hot"? Did Iraq have other nuclear facilities? Increasingly worried, Halim struggled to supply the men with answers to their endless questions. The fact was, Halim was too afraid not to. Something about these men was far too menacing for them to be simply businessmen. What had he gotten himself into? the Iraqi chided himself. A few weeks later Halim read in the French papers about the mysterious explosion at La Seyne-sur-Mer. Halim realized immediately that he had passed on the exact information about the date and place from where the cores would be shipped: information that the two Germans had been so interested in the previous month. The final straw came when the Germans asked, or rather demanded, that he introduce them to Yahia al-Meshad.

Meshad, an Egyptian-born senior nuclear physicist, had come to al-Tuwaitha from Alexandria University. He was in his mid-forties, dark and stocky, with a finely honed sense of irony and deadpan style of humor. By 1980, many of the young Iraqi scientists who had gone abroad to study were refusing to return to Iraq, especially the ones in the United States, where they began applying for asylum or appealing to the State Department. So Meshad, who had a reputation for brilliance and thoroughness, was considered quite a catch at the Nuclear Research Center, where he began working under Khidhir Hamza.

Hamza began to use Meshad as a liaison with the French engineers, sending him regularly to the Sarcelles nuclear plant to inspect the manufacturing of Iraqi reactor parts and equipment. When Chirac announced that France was going back on their deal and supplying Iraq with lower-grade caramelized uranium, Hamza assigned Meshad the task of ensuring that only the enriched $U235$ was sent to al-Tuwaitha. Weapons-grade uranium had to be 93 percent enriched. Caramelized uranium was enriched far below the enrichment required to extract plutonium. The substitute low-grade would be worthless to Iraq's atomic bomb program. Mossad was not certain how Hussein was going to respond to the new French dictum. Hofi, who opposed a military raid on Osirak, was especialy anxious to know if Iraq would accept the caramelized uranium. If so, then the threat posed by Osirak would be considerably less and perhaps Begin could be argued out of a military attack. The "Germans" were assigned to get the information out of Meshad—one way or another.

By now, Halim figured the German businessmen had to be spies. He prayed that Donovan could help him. After all, he knew these men. Halim telephoned Donovan in London and sheepishly relayed the entire story to him, including the part where he had been paid money behind Donovan's back.

Donovan feigned concern. According to Ostrovsky, Ran then set Halim up.

"I think Itsik and Goldstein may be CIA," Ran/Donovan replied.

"Oh, my God, they'll shoot me!" Halim panicked.

"No. No they won't," he assured him. "At least it's not the Israelis. The CIA won't hurt you. They just want information. Maybe I can help get you out of this. When are you meeting Meshad?"

"Tomorrow night," Halim replied. "For dinner."

"Good. I'll fly to Paris. Leave the name of the restaurant at my hotel. I'll stop by and pretend to run into you at your table."

The next evening Donovan happened by the bistro table. Halim invited him to sit down and introduced him to Meshad. He praised his friend profusely and, with an awkward, insinuating intimacy, explained to the physicist that Donovan was an amazing business-man who could buy and sell many things that were useful to them. He would pay generously for any "help" Meshad could supply. Cautious and arrogant, Meshad refused to take the bait. The din-ner ended with Donovan suggesting they could perhaps meet again. Meshad was noncommittal. Donovan watched him leave the restaurant. He was frustrated. He would have to depend on Halim.

Later at Meshad's hotel, Halim, anxious to satisfy Donovan and get the Germans off his back, again tried to recruit the Egyptian. Following Donovan's suggestion, he suggested the two have some fun and dialed Marie-Claude Magalle to join them. Meshad was perfectly open to Magalle's charms and retreated to his bedroom with her. But as far as doing business with the Germans, the Egyptian rebuffed all of Halim's approaches.

He was forced to call Donovan the following morning and give him the bad news.

First the cores had been sabotaged, and now he was supposed to deliver Meshad. Halim's head whirled. He actually had heart palpi-tations, he was so frightened. For the next month he tried to avoid the Germans and Donovan. He just wanted out. Finally in June, his wife, Samira, returned to Paris. Halim immediately confessed to her everything he had been doing.

"I think they are CIA," he told his wife.

"You fool!" Samira cried. "What do the Americans care about any of this? You have been duped by Israelis!"

Stunned and nauseous with the sudden epiphany, Halim realized his wife was right. In the days following her return, he discreetly wrapped up his affairs, packed up all their belongings, and bought two one-way plane tickets to Baghdad. By the end of June he and his wife were on a plane from Orly, Halim forgetting in his haste to say good-bye to Donovan.

Donovan, however, was unconcerned. Mossad had gotten out of Halim what they needed. The focus now switched to Meshad. Since Meshad was an administrator under Khidhir Hamza in Atomic Energy, the Israelis were convinced he was an integral part of Iraq's secret atomic bomb program as well. Mossad decided that if Meshad could not be recruited, *other* arrangements would have to be made—they might even have to show him a "better world," the *katsa* euphemism for an assassination.

In June 1980, Meshad returned to Paris to check on equipment and ensure that the uranium France was obligated to ship to al-Tuwaitha was enriched 93 percent. After a week at Sarcelles and a detour to the French countryside, he returned to Room 9041 at his favorite hotel, the Meridien, at around seven o'clock in the evening on Friday, June 13. Late the next morning a housekeeper again passed the DO NOT DISTURB sign that had been hanging on the room's doorknob throughout her entire morning shift. Anxious to clean the room and finish the floor, she slipped the key quietly into the lock and pushed the door open a crack, calling out, *"Allo, Allo."* Stepping in, she spotted Meshad's body lying on the floor beside the bed in a pool of blood, his throat slit.

The French papers in the following days reported that a hooker had propositioned Meshad in the elevator on his way to his room. Later on, the woman told inspectors she had heard men's voices as she stood outside the door to Meshad's room, though it was not clear if she had been asked by the Iraqi to come back later in the evening. The police concluded it was a professional job: some-one—a business partner, a competitor, or even a foreign intelli-gence agent—wanted to get their hands on some papers in the Iraqi scientist's room. This person, or group, had hired the prosti-tute to confront him outside his chamber and delay him. But Meshad had refused the proposition, then walked in on the perpe-trators and been killed.

The truth, though, was very different. Meshad, in fact, had made a date with Marie-Claude Magalle that night. He had been seeing the prostitute on every visit to France since the night Halim had first introduced him to her months earlier. Mossad, who had tapped Meshad's phone, knew that the Iraqi scientist had a date with the high-priced call girl later that night at the Meridien. Ac-cording to Ostrovsky, before Magalle arrived, an Arabic-speaking *katsa* named Yehuda Gil had knocked on the door to Meshad's room. The physicist cracked the door, leaving it chained. Gil qui-etly informed him that he had been sent from a "power" that would pay "a lot of money" for some information concerning the scien-tist's work for Iraq's Nuclear Research Center. Outraged, Meshad swore at Gil and told him to leave before he called hotel security. Gil, who was instructed only to make the offer, discreetly left the hotel.

Within minutes Marie-Claude arrived at Meshad's room. Indeed, she may have overheard Gil and Meshad arguing at his doorway. But it did little to spoil the Iraqi scientist's evening. The two had sex—Meshad, it turned out, had a weakness for S&M—and later that night Marie-Claude left Meshad sleeping peacefully in his hotel room bed. Shortly thereafter, a team of Israeli *kidon,* trained assassins, used a duplicated hotel passkey to silently slip into the scientist's suite. Without fanfare, they slit Meshad's throat and left him on the floor of his room, his life running out in a puddle beneath him. The prostitute, Magalle, was not in on the Mossad scheme. In fact, she did not even know that the mysterious men who called her were Israeli secret service. They were simply men who paid her generously to service customers they assigned her to and to provide information about the men afterward. She had her suspicions about who her mysterious employers might be, but what they asked for seemed harmless enough— mostly where her johns went, what they liked, whom they met or talked to, what they said about their jobs or personal lives, things like that. In her line of work it did not pay to ask questions.

But Marie-Claude was shocked when she heard about Meshad's murder the next day from another professional. She was frightened. For one thing, she worried whether the authorities would try to blame her. And might the people who killed Meshad come after her next? Panicked, she called the Paris police. The investigating officers interviewed her. Magalle told the inspectors about the male voices she had overheard. Not sure what to make of her story, the police took her passport and restricted her to Paris as a material witness. Several weeks later the lead inspector contacted Magalle and instructed her to come to police headquarters on July 12 for a follow-up interview. Two days before her appointment, on the evening of July 10, Marie-Claude was working a busy corner on Boulevard St-Germain on the Rive Gauche. A black Mercedes pulled up across the street. The man inside beckoned to her. As she crossed the street, another black Mercedes jumped out from the curb and, racing down the boulevard, ran straight into the hooker, sending her careening off the hood. She was dead before she hit the asphalt. In a scenario reminiscent of the mysterious black car that had sideswiped the pretty young girl outside the shipping warehouse at La Seyne-sur-Mer months earlier, the Mercedes and the driver were never seen again. Witnesses could not even remember clearly in what direction the car had headed after striking Magalle. Meshad's murder would go unsolved, if not unforgotten.

The French, however, soon had new mysteries to investigate. Before he had been sent to prison, Khidhir Hamza's onetime boss, Jaffar Jaffar, had begun pursuing a two-track process to create fissionable, bomb-grade uranium: while construction on Osirak was being completed, Jaffar would also begin work on the method pioneered by the Americans working on "Little Boy," one of the atomic bombs developed during World War II—magnetic enrichment. This alternate enrichment process uses huge electromagnets to separate uranium isotopes, known as EMIS for electromagnetic isotope separation, inundating the U235 with radioactive neutrons, making it suitable for bomb-grade fuel.

Salman Rashid, a bright, energetic young electrical engineer who had studied in Britain, was recruited to work with Jaffar on designing a huge electromagnet for the uranium enrichment. It was a tough assignment since much of the research was still classified in the West. Making things worse, neither Jaffar nor Rashid was particularly strong in mathematics, and the two encountered a good deal of trouble handling the very involved design calculations. The Iraqi NRC contracted with a Swedish company in Geneva, Brown Boveri, to help Rashid with the design.

About a month after Meshad was killed, Rashid set off for Geneva for a two-month research fellowship, accompanied by an Iraqi security officer and a half dozen assistants. It soon became clear to everyone involved in nuclear research in Geneva that Rashid was interested solely in magnetic enrichment. The week before he was due to return, the young electrical engineer suddenly came down with the flu. It was a particularly virulent case. Rashid was having trouble swallowing, and soon he began to bloat, his neck and jowls becoming alarmingly swollen. He was admitted to the American Hospital in Geneva, but the staff physicians were stumped. An Iraqi doctor was summoned, but he, too, could not identify the virus. No one had seen this kind of flu before. Six days after experiencing the first symptoms, Rashid was dead. An autopsy seemed to point to some kind of poisoning, though the exact agent could not be isolated.

The security officer insisted Rashid was never out of his sight, but colleagues admitted that the young scientist had frequented the many bars and restaurants in Geneva. Plenty of opportunities to poison or infect him had existed if someone so desired.

Several weeks later yet another Iraqi engineer, Abdul-Rahman Abdul Rassoul, visiting Paris on Atomic Energy business, was suddenly taken ill after contracting "food poisoning" at an official French banquet. He died within days.

The epidemic of Iraqi nuclear scientist deaths and the sabotage at La Seyne-sur-Mer continued to raise eyebrows in France, a country never at a loss for conspiracy theories to begin with. By summer the French media were full of speculation about who was behind the mysterious attacks. And more than ever, France's complicity in Iraq's nuclear program was continually in the news. The head of France's Nuclear Energy Commission, André Giraud, a longtime critic of the treaty, warned that Iraq could well be seeking nuclear weapons. Others in the scientific and military communities were alarmed as well. One American nuclear expert pointed out that a reactor such as Osirak was designed for "nations engaged in the indigenous production of nuclear-power reactors. Iraq would have no great economic or energy incentive to establish a nuclear power generating capacity." What did that leave other than military purposes?

Chirac insisted that the French Atomic Energy Commission was in complete control of the reactor. He had already announced that France would supply Iraq with only the caramelized uranium. But Hussein, it turned out, was having none of that. He began a counter media campaign, deriding the Western nations that were now so afraid of Iraq.

"These Arabs, the Zionists said, could do nothing but ride camels," Hussein scoffed sarcastically. "How could a people who only know how to ride camels produce an atomic bomb?"

Iraq invoked the treaty: the nation would settle for nothing less than the original deal: seventy-two pounds of 93-percent-enriched weapons-grade uranium. Any substitute, they insisted, would not allow Iraqi scientists the full range of "peaceful" research activities planned at al-Tuwaitha. Changing the deal would force Iraq to suspend payments—and perhaps even the oil shipments—negotiated in the original treaty. France quietly decided to abide by the original agreements: Iraq would get its seventy-two pounds of fully enriched uranium. But Chirac insisted that French scientists were still in charge.

From the earliest days, when he began to plan a possible mission to strike Osirak, General Ivry worried that the reactor would go hot before he could hit it. Once the reactor was radioactive, any bombing would bring the risk of fallout and large civilian casualties, potentially in the thousands. Despite Hofi's operations, Iraq had continued steaming ahead undaunted. The cores that Ivry had initially hoped to see destroyed at La Seyne-sur-Mer had been repaired and shipped to Baghdad. He knew about the deaths of Meshad and the others, even though Hofi, of course, never spoke of such things openly. But the deaths of a few nuclear scientists, while creating chaos and anxiety inside al-Tuwaitha, did little to stop Iraq's mammoth nuclear program.

Recent intelligence by AMAN and Mossad reported that Isis, the smaller French "companion" reactor to Osirak, had been completed and was already hot. French scientists working at al-Tuwaitha interviewed by undercover Mossad *katsas* indicated that the main reactor would be fueled by July 1981—less than eighteen months away. Israel's scientific experts predicted that Iraq could have an atomic bomb by 1985.

Ivry could almost feel the days peeling like fallen leaves off the calendar on the wall behind him, whittling down to the final day when Osiris would become hot and the game would be over. Hussein would soon after have the ability to destroy Tel Aviv, Haifa, and Jerusalem, literally with the flick of a switch, as the cliché went. And still Ivry had no approved plan and no green light from the cabinet to forestall this nightmare.

The general, working at his headquarters with a small, hand-picked staff, had already discarded a half dozen plans. At first he had considered putting an insertion team into Iraq, using a combination of air transport and assault tactics similar to the commando raid at Uganda's Entebbe Airport in 1976 that had freed the Palestinians' hostages with a minimum of casualties. But there were too many negatives. For starters, maintaining longtime secrecy for such a complex mission, employing hundreds of men and multiple assets from all military branches, would be almost impossible. Next, the logistics were daunting. Transport planes would have no place to wait on-station during the operation. Flying six hundred miles to Baghdad, dropping off the assault team, returning to friendly borders to wait, and then flying back to Baghdad *again* for the extraction was obviously unworkable. Besides, the transport planes would need to be refueled, and the delicate operation of refueling over hostile territory in the unpredictable turbulence of hot desert air hundreds of miles from home was out of the question. Photographs showed that the Iraqis had reinforced the periphery of al-Tuwaitha with twenty-foot-high earthen revetments, ringed at the top by concrete and electrified wire fencing and antiaircraft gun towers, making a successful storming of the fences unlikely. Finally, the probability of men being captured in a ground assault was high, and Ivry refused to risk the certain barbaric treatment of his men at the hands of Iraqi security thugs.

Ivry's decision was later validated once and for all with the fiasco of the U.S. rescue mission to free sixty-six embassy hostages in Tehran in 1980. The Delta Force raid was a night extraction, the teams flown by transport to a makeshift landing strip in the middle of the desert miles from the capital. Unfortunately, a sandstorm blew in out of nowhere—not uncommon in the region—blanketing the troops and throwing everything into confusion. Unable to maneuver in the heavy winds and zero visibility, one of the army choppers swept sideways into a plane and exploded, killing six commandos and the pilots immediately. The mission was aborted, the U.S. humbled and humiliated. Back in the States the disaster came to symbolize President Carter's weakness. To Ivry, it was just a hard lesson learned.

"Too many things can go wrong," he told Eitan.

Month after month Ivry and his staff computed and modeled, calculated and experimented, guessed and second-guessed. Again and again he found himself returning to an air attack. But there were many problems with such a mission as well. The first hurdle was deciding what planes to use. As Ivry and his command staff studied the options, it became apparent that each fighter in the IAF's considerable arsenal carried a serious negative.

Popular since the Yom Kippur War, the single-engine A-4 Skyhawk and the Israeli-built Kfir were the country's primary attack fighters. But both aircraft lacked sophisticated new radar and bombing systems, and neither had the range round-trip without refueling. The F-4 Phantom, a heavy-duty, twenty-year combat veteran that the United States had continually upgraded with modern equipment, carried PMGs, precision-guided missiles, an early-generation "smart" weapon whose accurate targeting was controlled from inside the plane and might be needed for such a raid. But the Phantom was bulky, hard to maneuver, and a gas guzzler. Moreover, it flew with a two-man crew, doubling the number of men who could be killed or captured with each plane.

The newest and most sophisticated of the fighters was the American F-15, with twin Pratt & Whitney F-100 engines and a look-down, shoot-down pulse Doppler radar, a highly sophisticated computerized radar that functioned by sending out continuous pulsing signals, which allowed pilots to lock on targets flying as low as twenty feet off the ground and still distinguish them from the "noise," the confusing ambient signals emanating from the terrain below—a problem with older radar systems. Designed as an air-to-air fighter, the F-15 held the world record for altitude speed climb and had a lock-on target range of one hundred miles, the best in the world. General Avihu Ben-Nun, the northern Israel IAF commander at Tel-Nof Air Base, home of the F-15 wing and Israel's top-secret nuclear-weapons-capable squadron, lobbied hard for using his planes, arguing that they were the IAF's most sophisticated asset. Of course, such an important assignment would also confer great prestige and influence on Tel-Nof and Ben-Nun.

But Ivry held off. Some on Ivry's staff had serious doubts about whether eight F-15s could make the trip to Baghdad and back safely. For one thing, the F-15 engines had not proved reliable and had a high rate of mechanical failure and maintenance. Most damaging, the fighter did not have the range to fly round-trip without CFTs, special conformal fuel tanks that could be bolted to the sides of the fuselage above the wings, doubling the distance the plane could fly. The United States, under pressure from the Soviet Union and Arab states to rein in Israel's military and preserve the balance of power in the Middle East, refused to sell Israel the conformal tanks as part of its arms control cutback.

As the months went by, Ben-Nun grew more and more frustrated with what he considered Ivry's foot-dragging. He argued that the Tel-Nof wing would have the CFTs by the time of the raid. Israel was already lobbying the U.S. Departments of State and Defense to amend the agreement and sell Israel the tanks. In addition, the Israeli defense industry had begun exploring the possibility of manufacturing its own external tanks. But Ivry worried about losing pilots, and the F-15 increased that risk.

Then, on February 1, 1979, events a thousand miles to the northeast literally changed Ivry's world. A month after Jimmy Carter had toasted the shah of Iran, Mohammed Reza Pahlavi, for maintaining an "island of stability" in his troubled corner of the world, the hollow-eyed, exiled Ayatollah Ruhollah Khomeini landed in Tehran with a chartered plane full of fanatical fundamentalist followers from Paris and proclaimed the Islamic revolution. Overnight the balance of power in the Middle East was reshuffled. For Ivry, it turned out to be a textbook example of the Law of Unintended Consequences.

Some weeks after the headlines had shaken statesmen and military planners from the Knesset to the Capitol, Defense Minister Ezer Weizman rang Ivry in his Tel Aviv headquarters. The U.S. Defense Department had just called, Weizman informed him. It seemed that the shah of Iran had negotiated a contract to purchase 160 of America's new high-tech F-16 Fighting Falcon combat jets. The United States, not surprisingly, had canceled the contract following the revolution. General Dynamics Corporation, the plane's San Diego–based manufacturer, was stuck with seventy-six of the aircraft scheduled for the production line, with the first eight already being assembled. Would Israel, Defense wanted to know, be interested in purchasing them?

Ivry could practically *hear* Weizman smiling to himself on the other end of the line.

"*Ken!*" Ivry said.

"Yes." He would be happy to help the United States out. After all, what were friends for?

Conceived as a faster, smaller, lighter complement to the F-15, the single-engine F-16 went into experimental design and production at General Dynamics in 1975. The plane first flew in December 1976. Almost futuristic, designed for a single crewman, the fighter was the first aircraft to employ revolutionary space-age technology and manufacturing. Many of its parts were made of special composite materials that did not reflect radar beams and made tracking difficult, forerunners of the engineering that would eventually make possible the "invisible" stealth bomber.

Instead of the traditional mechanical control stick between the pilot's legs, the F-16 operated by FBW, or fly by wire, a computerized control system guided by a pressure-sensitive handle on the right side of the plane that more resembled a Nintendo game stick. The handle sent electrical impulses from the control surfaces to a computer, which guided airspeed, lift, banking, sharp turns, dives—virtually all mechanical controls. A new computerized BITS, built-in test system, checked out the plane's mechanical, electrical, navigational, communications, and weapons systems in seconds before takeoff—a process that would take techs and pilots in older fighters up to fifteen minutes, and still not cover a third of the internal mechanics BITS monitored.

The newly invented, *Star Wars*–like HUD, heads-up display, showed readouts on a see-through glass screen mounted at the pilot's eye level so that during combat he could check G-loading, airspeed, mach number, altitude, and time- and distance-to-target readouts without ever having to take his eyes off the sky. He could also "click on" weapons icons to choose air-to-air missiles, machine guns, or bombs. The turning radius of the F-16 was one-half that

of the F-4 and far better than any MiG's. Such acute, radical turning tremendously increased the G forces pulling at the pilot. Fliers
would normally black out at six or seven Gs, even wearing a pres-
surized G-suit. The cockpit seat in the F-16, however, was designed at a 30-degree tilt so that the pilot's feet and buttocks were
on the same level, making it harder for the pilot's blood to leave the
head and be pulled by negative Gs to the body's extremities. The
tilted seat allowed pilots to move with the plane and remain conscious at as much as nine Gs for short periods of time. And because the pilot sat in a glass bubble canopy nearly level with the
fuselage (the origin of the plane's derogatory nickname, the "glass
coffin"), he could scan virtually 360 degrees and easily "check six,"
that is, look all the way around for enemy planes on his tail.

Bombing, too, was computerized. The weapons system employed a graphic bomb line shown on the screen that locked on the
target, allowing the pilot to guide the aircraft along the line until a
circle at the end, called the "death dot," covered the target. The pilot then pressed the pipper, the "bombs-away" button, and "pickled
off" the bombs, which could hit within 15 feet of the target. The
F-4 hit within 150 feet.

Ivry was convinced the F-16 could give his pilots a much
needed technological advantage in attacking Osirak—if the plane
performed in actual combat as advertised. But questions remained: When could the fighters be delivered? And how could a
radically new aircraft be integrated virtually overnight into IAF
doctrine, one of the most complex in the world?

Known as Cheyl Ha Avir, or Corps of the Air, the IAF's dogma
and tactics were highly sophisticated, continually being refined by
combat experience. By 1980 the IAF had recorded some 700 kills
in the air, and 13,000 total since its inception in 1947—more than
all fighter kills by all the countries in World War II combined.
Planes were maintained 24–7. Pilots were drilled mercilessly in

the assets they had. Given the present timing of Osirak's completion, to even consider using F-16s against the reactor, the pilots, support crews, and maintenance techs would have to be trained and "expert" in the plane's electronics, weapons, and navigation systems in six months—a process that normally took two years.

To pull it off, Ivry would need the best and most fearless pilots. And some, perhaps, a little crazy.

In the fall of '79, the general asked his IAF commanders to recommend their best pilots to begin training in the F-16s. Each commander argued for his nominee, but ultimately Ivry personally selected twelve pilots to attend the USAF's Operational Training School for the F-16 at Hill Air Force Base outside Salt Lake City, Utah.

The assignment was considered a huge honor and was passed down the normal chain of command. The operational training regimen lasted ninety days. The men would attend in three groups of four, then return for reassignment to Ramat David Air Force Base, north of Tel Aviv in the heart of the Jezreel Valley. Ramat David would be home base to the IAF's newly constituted F-16 squadron, under the direct command of the base commander, Col. Iftach Spector, a nearly legendary F-4 Phantom pilot in the IAF. Though the ultimate business plan of the new franchise would still have to be worked out, for now its first mission would be to form two squadrons expert in the F-16 and ready to train other pilots. No one outside Begin's security cabinet and IDF high command knew anything about Osirak.

The first pilot Ivry called to headquarters was Zeev Raz, a thirty-two-year-old lieutenant colonel and father of four. Zeevi, as he was known by his pilots, was a wing commander. Like the general, Raz preferred to skip the b.s. Trim, athletic, with dark hair, he had warm, friendly green eyes that helped soften his brusque, all-business demeanor. In his preoccupation to just get the job done, Raz could be rude, but he was also understanding and loyal, loved to read and study history, and doted on his four children. Neither man had time for the "interior" life. Asked what they "felt" about something, each would look puzzled. Ivry would screw up his wispy mustache and raise his eyebrows, as though such an idea had never occurred to him. In life, there were things a man wanted to do, things a man needed to do, and things a man *had* to do. What did feelings have to do with anything?

Raz had dreamed of flying ever since he was a small boy living in the tight-knit farming community of Kibbutz Giva in the Jezreel Valley. He had been bitten by the flying bug the day of his bar mitzvah, having been given a telescope as a present. As the guests danced and ate, Raz sneaked out to the backyard and used the telescope to watch a squadron of French Mirages land at the air force base just a half mile from his home. They seemed so beautiful and graceful. And how wonderful to soar far above the earth, free of its cares and problems. But the truth was that, despite his dreams, deep down, Raz, a sensitive, introverted boy who loved to read and think by himself, could never truly see himself as a fighter pilot. So after high school, when he asked to test to enter elite pilots' training, he was stunned to learn that he had actually passed the exams. Indeed, after he enrolled in flight school, he found that his instincts had been correct: he was not a "natural" flier, the way some people were born musicians or athletes. Learning to fly, to become an Israeli fighter pilot, one of the world's most elite warriors, proved almost impossibly difficult.

Enrolling at eighteen or nineteen right out of high school, instead of following college as in the United States, only two out of ten candidates made the cut at Hatzerim Air Base in southern Israel. But fiercely dedicated, obsessively thorough, Raz fought his way through to become one of the IAF's top pilots. By the time he graduated to F-4s, he had proven to be a natural leader as well.

On the afternoon of October 7, 1973, Raz would finally have the chance to do what he had dreamed of all his life: lead a squadron of brave men into a desperate battle. The preceding afternoon, at exactly two o'clock on the Sabbath of Yom Kippur, the holiest day on the Jewish calendar, Syria sent the most artillery, tanks, and infantry arrayed since World War II pouring over the Green Line, sweeping down the Golan, and nearly overrunning the fragmented IDF brigade assigned to hold the north. Subjected to a murderous counterbattery fire, outgunned 12-to-1, the Israeli mechanized artillery brigade tried desperately to hold the line. The Israeli Air Force had the fastest turnaround time from takeoff to takeoff of any nation on earth. But the sorties of F-4 Phantoms and Skyhawks cycling through Beersheba to provide air cover to the besieged divisions were being decimated by new radar-controlled Soviet SAM-6 surface-to-air missiles. Unbelievably, the Syrians were on the brink of breaking through and racing to the coast, cutting Israel in two.

Raz led his squadron off the runway, climbing steeply to the northeast, the sun a bright ball in the west behind him. The F-4s headed straight to the Syrian border. Over Mount Ramon, Raz spotted his first MiG-21. The MiG turned to engage, and Raz fired off an air-to-air Sparrow. Seconds later the MiG exploded in flame and smoke, and Raz had recorded his first kill. He felt a rush of adrenaline, but he took no joy in the death of the Syrian pilot. It was a job, something he had been trained to do—and do expertly. Within minutes the squadron chased out the Syrian MiGs and provided air cover for the besieged battalions battling below. Twenty-four hours later, Israel had managed to reinforce the northern battalions and halt the Syrian advance.

Raz's discipline, attention to detail, and ability to organize and lead men lifted him steadily up the ranks of the IAF, first as an F-4 squadron leader and then as one of the renowned instructors at the IAF pilots school at Hatzerim. From the beginning, Ivry picked him to be his F-16 leader.

Ivry greeted Raz warmly after the pilot transferred to Ramat David. They made small talk, Ivry asking about his family. With its history of volunteerism, the Israeli military was more informal than most, especially compared to that of the United States. Not a lot of stock was put in marching and drilling and saluting.

Ivry got down to business quickly. He wanted Raz to lead the first group to Hill. "Who else is going?" Raz asked.

"Hagai Katz, Relik Shafir, and Doobi Yaffe," Ivry said.

Raz nodded. Dov "Doobi" Yaffe was a good man. It would be good to work with him again.

The two pilots had attended the U.S. Navy's famed "Top Gun" school together at Miramar, California, just east of San Diego. Ostensibly enrolled to hone their air-to-air combat skills, Raz and Yaffe were in reality sent to collect intelligence on the new F-5e jet fighters. The Saudis and Jordanians had bought scores of the new American-made planes to beef up their air forces, and Israel was worried about the plane's performance specs and capabilities in combat. Top Gun was one of the few places that flew F-5es to simulate "enemy" aircraft during training exercises. By enrolling, Raz and Yaffe could study the planes in action. The Americans, of course, were not let in on this secret.

Yaffe and Raz had at first been uneasy together at Miramar. They had not known each other in Israel. And they were polar opposites. Israel then was perhaps the closest thing the industrial world had to a "classless" society. Even money did not buy you nobility. There were no Rockefeller or Kennedy Israelis. But that was not to say there were no social divisions. In *Eretz Yisrael* the social pecking order was based on how close your family tree grew to the nation's heroic founding fathers and the generation of young men who had fought for Israel's independence. In this sense, Doobi Yaffe was practically royalty. His grandfather Dov Yaffe had come to Israel as a Zionist pioneer from Russia at the turn of the century and founded the first agricultural settlement in Galilee. His father, Avraham, was one of the first Israeli fighter pilots and commander of Israel's largest air force base, Tel-Nof. Doobi's uncle "Yossi" was a national hero, a tough paratrooper who led the famous charge in the Battle of Ammunition Hill, known to every Israeli schoolboy and -girl and immortalized in the song *"Givat Hatakh Moshem."*

On the second day of the Six-Day War, Jordan's crack 2nd al-Husseini Battalion held the hill, blocking the entrance to Jerusalem. Named after the garrison where the British had stored their ordnance during World War I, the Ammunition Hill fortification was a labyrinth of minefields, trenches, and concrete bunkers manned by artillery and tanks. Six barriers of barbed wire ringed the hill. Yossi Yaffe had to get his company up the hill, through the barbed wire, and then seize the concrete bunkers. Chewed up in a no-man's-land by blistering machine gun fire, the paratroopers crawled barrier by barrier, using pipelike Bangalore torpedoes to destroy the barbed wire. When they finally reached the hill, the paratroopers fought hand to hand in the bloodiest combat of the war. At 03:10, Yossi radioed Commander Motta Gur that the hill had been taken. The commander shouted back: "I could kiss you!" Yossi had lost a quarter of his command, but he had opened the way for the historic taking of Jerusalem and the Old City. When he was killed by a land mine a decade later, half the nation turned out for his funeral.

Raz was aware of Yaffe's pedigree. He seemed the well-connected class smart-ass, good-looking, easy-to-know. Raz on the other hand was the no-nonsense, mission-oriented, quintessential team captain, the up-by-the-bootstraps kibbutz boy. But over the months the two men drew closer together. For one thing Yaffe was no prima donna. A jokester, he could always get a laugh out of Raz, and brought him into the group of American pilots he had befriended. The two would have an occasional beer at local off-base clubs with the Yank pilots. Yaffe appreciated Raz's knowledge of flying and his attention to detail. In turn, Raz deeply respected Yaffe's professionalism and courage, and especially his incredible, God-given talent to fly. Yaffe was a born ace. By the time they left Miramar, the men were fast friends.

On a whim, before returning to Israel, Raz and Yaffe took a side trip to General Dynamics' manufacturing plant in nearby San Diego, where they were given a tour of the production line, which was busy assembling the aerospace firm's brand-new, cutting-edge aircraft, the F-16. The idea that they might one day be asked to fly the plane never occurred to Raz or Yaffe.

When Ivry informed Yaffe of the assignment, he could not believe his luck. Unlike the skeptical old veterans, Yaffe had been enamored with the F-16 since the trip to General Dynamics. Raz's team was dispatched to Hill in early February 1980, while Ivry finished assembling the second team to leave in May. For team leader he chose Lt. Col. Amir Nachumi.

Tall, boyishly handsome, with sandy brown hair and thin as a reed from a Jordan River bank, Nachumi was a bundle of energy. As far back as high school in the early sixties, he and his classmates had bet on who would get into the toughest army unit. At graduation, barely eighteen, he immediately applied to the air force's notoriously torturous pilot training. But after passing the entrance exams, Nachumi had stumbled during a desert survival test, succumbing to the heat and fainting briefly. During the medical checkup, the air force doctor, a major, pulled a crisp new twenty-shekel note from his pocket and scraped it across either side of the young recruit's pink cheeks. Satisfied, the doctor nodded to himself and, stuffing the bill back into his pocket, sent Nachumi on his way, telling him at the door: "Come back when you start shaving."

Instead, Nachumi joined the tank corps. Following compulsory service, he attended Hebrew University in Jerusalem and, two days before graduation, was called up as a reserve tank officer at the outset of the Six-Day War. After a murderous week of fighting in the Sinai, driving Nasser's mechanized units back to the canal over the craggy desert terrain, Nachumi, dusty and sweaty, looked up one day from the furnace of his turret to see a wing of shiny Israeli Phantoms strafing gracefully across the cool blue skies above and wondered, What am I doing *down here*?

When hostilities ceased, he finished school and then, more determined than ever, reapplied to the IAF. The first man he met was the commander of the air force flying school at Hatzerim, Col. David Ivry. Looking up from Nachumi's military file, Ivry stared at the young tank commander.

"Why are you here?" he asked.

"I was stopped by some of your doctors," Nachumi replied. "But now I'm back."

By the time Nachumi finished his story, Ivry had approved him on the spot.

Five years later he was a twenty-eight-year-old F-4 Phantom pilot stationed in the wilderness of Ophir Air Force Base at Sharm al-Sheikh on the sweltering tip of the southern Sinai Peninsula, when unconfirmed reports came in that Israel was being attacked. Unlike Raz in the north of Israel, Nachumi had no way of knowing that Egypt had already crossed the banks of the Suez, the "canal of shame," as President Anwar Sadat had called it. But he knew that if a war *had* begun, Egypt's first target would be Sharm al-Sheikh, the long-disputed gateway to the Red Sea lying in the V between the Suez and the Strait of Tiran.

Without orders, Nachumi rounded up his wingman and their weapons officers and sprinted to the runway, firing up the F-4's GE turbines. Stationed at the front, the planes were already fully armed. Despite orders over the cockpit radio to "delay," Nachumi and his wingman taxied their Phantoms down the runway and climbed steeply up, away from the base. Seconds later, as the two pilots circled to the west, they saw a squadron of Egyptian MiG-17s and 21s pounding the airfield below.

Nachumi's radio crackled. It was Ophir command. "Who is this? Who's the flight leader?" the base commander demanded.

"There is no flight leader," Nachumi replied. "Just the two of us."

There was a long silence.

"You're it then," came the reply.

Nachumi turned and dived, facing off against twelve Egyptian fighters. It was his first combat experience, and he was outnumbered six to one. He fixed the lead MiG on his radar and fired off his first missile. The MiG dived, trying to shake the heatseeking Sparrow. It failed, bursting into flames and plummeting to the desert floor far below, where it blossomed in a puff of smoke like a tiny blue mushroom. Nachumi, pulling up fast, lost an engine. The heavy plane began stalling out as he quickly refired the turbine. It caught and he resumed the attack. His wingman beside him arced left and launched another Sparrow, which dropped and acquired its target, flashing across the sky toward a second MiG-17. It disintegrated in a flash of fire and black smoke. In the end Nachumi shot down four MiGs and his wingman three. As the two Phantom pilots circled above the airfield, the remaining five MiGs turned and headed west back to Egypt.

Nachumi would later be awarded the OT HAOZ, the IAF's second-highest award for courage on the field of battle. No one was prouder than his former flying school commander, Brig. Gen. David Ivry.

Ten minutes after he walked into Ivry's office in February 1980, Nachumi "volunteered" to lead the second team to Hill.

By February, the Italian manufacturer SNIA Technit was finishing work on Iraq's chemical reprocessing unit and its main components—the "hot cells," shielded labs designed for handling radioactive materials in safety and for separating plutonium from the spent fuel. President Carter had personally asked Italy to reconsider selling Iraq the hot cells at the time the deal was discovered, but the Italians demurred. As a deal sweetener, Iraq had agreed in addition to the hot cells to also purchase four Italian naval frigates—despite the fact that they were powered by U.S.-made General Electric turbines. Italy assured the United States and Israel that they had nothing to worry about: it was sending its own technicians to al-Tuwaitha to guarantee the Iraqis would never use the equipment to separate weapons-grade plutonium from the spent fuel rods. Finally, the Israelis could sleep in peace . . .

One balmy Mediterranean evening in August, the getaway month for French and Italian workers, a bomb exploded on the front porch of the SNIA director's tony "villa" apartment in the suburbs just outside Rome, obliterating the front of the building. The director was out of town with the rest of Rome, and so was spared. Simultaneously, two other bombs exploded inside SNIA's Rome headquarters, causing extensive damage. The director rushed back to the capital to investigate and assess the damage. As with the Seyne-sur-Mer bombing, a group no one had ever heard of claimed responsibility for the blast the next day: the Committee to Safeguard the Islamic Revolution. This time, instead of a phone call to the city newspaper, a message was left for the SNIA director: "We know about your personal collaboration with the enemies of the Islamic revolution. All those who cooperate with our enemies will be *our* enemies." Demanding the company quit all business dealings with Iraq, the note went on to warn: "If you don't do this, we will strike out against you and your family without pity."

Hofi had not given up on buying Ivry more time.

The second lieutenant sat just inside the open doorway of the CH-53, the blades of the heavy-duty combat helicopter whirring overhead and kicking up dust from the asphalt. He and his six-man CSAR (combat search-and-rescue) team had sat steaming on the hot tarmac waiting to spin up for more than an hour, inhaling diesel fumes and trying to find a comfortable position in the back of the cramped chopper. His desert fatigues were already sweat-ringed under the arms and around the collar. At 1501 the mission chief appeared out of the operations hut and signaled the "go" sign. The lieutenant pulled his body back into the bird so that his legs cleared the door as the engines revved, the vibrations shuddering the metal frame and rattling his teeth until the chopper lifted off the tarmac and climbed skyward, its nose banking eastward as it gained speed, already crossing the rugged sands of the Negev toward the border. He ran through his orders: CSAR would helo east, maintaining radio silence, to the Jordanian border, then hold position just west of the line, hovering at one hundred feet. There, they would stand ready, awaiting word of downed pilots, the CH-53's tracking frequencies tuned to the mission pilots' PRCs. The CSAR team had been given no briefing about the mission or the pilots' destination. Their orders were simple: in the event of pilot downing, proceed east, violating Jordanian and any and all sovereign airspace as necessary, and effect rescue as quickly as possible, then return to base. All secrecy was maintained. In fact, as far as anyone outside the base was concerned, CSAR had not even been deployed.

the Golan while he remained behind. Now, finally, he was in The Club.

The twelve pilots selected to form the core of Israel's new F-16 Fighting Falcon squadron began moving their families into the officers' quarters at Ramat David in the late fall of '79. The residential area housed some one hundred families. The accommodations were small two-bedroom apartments, each with a living room and a tiny kitchen, located about twenty meters apart in neatly landscaped rows about three hundred meters from the base runway, so that the wives would hear the constant roar of their husbands' planes taking off and landing all day long. As a perk, the pilots, most of whom had families, had day nurses to aid their wives in the care of their children. It was not an uncommon sight to see smartly dressed nannies wheeling a small squadron of tykes down base streets in wooden strollers that looked more than anything like small wooden market baskets—or toy circus cages.

The squadron's first assignment was to become expert in the design and mechanics of the F-16's various systems. Flying tactics, or "switch orientation"—which buttons did what—would come later. Each pilot was assigned one of the plane's systems to study in detail, to master the theoretical principles behind its design and the mechanics that comprised its operation in order to teach future trainees. The pilots also translated the original General Dynamics specification and operation manuals from English into Hebrew and wrote up the IAF's own conversion courses to be used in training incoming Israeli F-16 pilots. Yaffe, who moved into quarters just down the road from Katz, was responsible for the engines. Hagai was made both weapons specialist as well as F-16 project manager, responsible for determining the IAF's requirements for the entire aircraft and then ordering the appropriate design and production modifications from General Dynamics in San Diego.

The process was complicated by the fact that the first eight

F-16s off the production line were Iranian and had been built to the shah's specifications. Israeli Air Force tactics and combat protocols were significantly more sophisticated than those of the Iranians, and therefore so were the IAF's operational needs. For one thing, the shah had ordered his planes built to United States Navy specifications. As such, the planes were designed to carry the navy's standard 1,000-pound ordnance. The IAF on the other hand used only 500-pound or 2,000-pound bombs. Katz had to ensure that the F-16s already on the line were retrofitted with the proper bomb clips and release systems to accommodate Israeli ordnance as well as make sure the necessary design changes were incorporated into production of future Israeli F-16s. The IAF's F-16s would be outfitted with Israeli-made, air-to-air Sidewinder missiles instead of U.S. Stingers. This change necessitated not only new modified missile clips under the wings but also changes in the plane's computerized missile logic. Likewise, the Iranian-ordered F-16 communication systems had to be replaced and modified to conform to the IAF's own secure Com links. There were many such changes throughout the aircraft's various systems.

Katz ordered the retrofittings and/or design changes through the GD liaison engineer, now posted to Tel Aviv. Some changes were relatively simple mechanical replacements; others would take months to design, build, and implement, delaying the delivery of the planes. The first two planes off the line would be F-16-003s, two-seaters that would allow OTU pilots to train alongside future conversion fliers. The bulk of the F-16s, however, would be single-pilot fighters. By December 1979, Katz finished his project work overseeing the mechanical conversions and began to think about something other than the insides of the plane. In two months he would ship out with Zeev Raz and the first conversion team to Hill Air Force Base. He was excited about the prospect of flying again. That was what it was all about, after all.

The first of February 1980, Raz flew ahead to Hill outside Salt Lake City. The rest of the squad, Yaffe, Katz, and a young F-15 fighter pilot named Israel Shafir, followed later in the week, flying a commercial El Al flight from Israel to New York, then transferring for the three-hour connection to Salt Lake City. The pilots were allowed to bring their families with them to Hill for the three-month training course. Yaffe brought along his wife, Michal, and his young son. Both Shafir and Katz brought their wives as well. Raz, intensely focused as leader and commander, rode alone.

The pilots and their families moved into a small motel in the tiny military town that had grown up outside the base, which was situated in a valley beneath the Rocky Mountains. Raised in the deserts of the Middle East or the flat fertile plains off the Mediterranean, most of the Israelis had never seen anything like the towering, snow-capped peaks that surrounded them. The pilots were awestruck by the majestic backdrop of the white-topped ridges that rose up to encircle them as they descended to the runway below.

The pilots attended flight school five days a week and had the weekends off. Indeed, the conversion training was more like a nine-to-five job than Hollywood's *Top Gun* camp—with the one crucial exception that their workday could well end with their funeral. The pilots were professionals, most far along in their careers. They resembled more a steeled special ops team with a job to do than the clichéd cocky cowboy-pilot. Unlike infantry combat units where men often did form tight friendships or fraternities, single-seat fighter pilots were loners. The feeling was more or less that you were on your own. You did the flying alone; you made the crucial decisions alone; if you were hit, you died alone. The camaraderie between fighting men expressed itself in joking or good-natured hazing. Feelings, fears, even missions were never discussed.

But despite this natural proclivity for solitude, the men and their families, working and living together so closely, especially so far from home in a wholly alien environment, formed a close-knit community, a little Israel of the Rockies. Though not overly religious, the men and their wives felt it important to keep up family traditions and the culture of their homeland. So every Friday evening all of the families, including Raz, would gather at the motel home of one of pilots and celebrate the Sabbath, lighting the candles, singing folk songs, and sharing favorite dishes. Yaffe, wisecracking, easygoing, even irreverent, kept everyone loose. Shafir, a superb musician, played guitar and led the songs.

Known to everyone as "Relik" or "the Major," Israel Shafir was the unofficial squadron philosopher. Dark-haired, with intense brown eyes and classic Roman looks, he was educated, urbane, well spoken. He liked to read historical novels and write poetry, though he didn't like anyone to know that. Shafir had lived and gone to school in Canada and Scotland, spoke perfect English, and evinced a devilish, self-deprecating humor, routinely addressing everyone in mock aristocratic accent as "Doctor." If asked a question by one of the OTU instructors, Shafir typically answered playfully à la Cary Grant, "Yes, Doctor?" or, when walking across the tarmac, would call out in greeting, "How are you today, Doctor?" When he finished with flying, he planned to return to university for a doctorate in philosophy.

Some nights the Israelis would join their Yank instructors at the local base bar for a round of beers, some good-natured nationalist boasting, and even a song or two. On weekends the Israelis turned out like tourists, traveling to national parks Bryce Canyon and Zion. In an "only in America" moment, they caught the local tour of a demolition derby, watching crazy Yanks smashing perfectly good cars into one another. Later, during training for the third group, the pilots and their families were shuttled to Park City's famed ski resort. Dressed in T-shirts and Levi's that were standard wear in the Middle East, the Israelis were stunned to discover a picture postcard forest of snow packed ten feet deep. Bare-armed and freezing, they were like kids and spent the day hiking the drifts and challenging the instructors to snowball fights.

But mostly the pilots were excited about flying the new machines, the F-16s. Raz's group in March eagerly awaited delivery of Israel's first four aircraft to Hill—the two two-seaters and two single-seaters. Finally the day came when the four aircraft, painted virgin military gray, without wing and tail insignia, sat quietly, regally, on the tarmac before the Hill hangars. Later the planes would be emblazoned with the familiar blue circle and Star of David of the Israeli Air Force, but for now, for the pilots, the planes on the runway, sparkling in the early morning sunlight, were tangible symbols of their nation's strength. They were inexplicably proud. And excited as children at Christmas—or, at least, Hanukkah. The planes would ultimately be flown to Israel, but for now they were the aircraft Raz and his pilots would fly during training. No longer would they need to borrow the USAF F-16s for conversion exercises.

Yaffe had studied the design and mechanics of the plane for months, but even so, the first time he climbed into the cockpit of the multimillion-dollar, state-of-the-art fighter, he was nearly overwhelmed. The sensation of sitting high in the glass-canopied cockpit, tilted at a 30-degree angle and raised above the lines of the fuselage with nothing but blue sky surrounding him, was stunning. The powerful Pratt & Whitney engine, responding to the lightest touch of the computerized control stick, would literally blast him forward, pinning Yaffe to his seat like a race car driver coming out of the pits. For Yaffe, in fact, it was like being in a brand-new Porsche.

Training under command of the 16th Tactical Fighter Training Squadron was a mix of classroom, tactical flying, debriefing, and instruction that allowed the pilots to grow accustomed to the new aircraft. The course was designed to evolve through a graduated

series of training flights and instruction, including intensive training in BFMs, basic flying maneuvers, emphasizing one-on-one
dogfights between the pilot and an OT (operational training) instructor; ACMs, air combat maneuvers, learning combat tactics
with two bandits against one "good" guy; and ACTs, air combat
tactics, concentrating on defensive and offensive maneuvers against
a group of bandits. The pilots also flew training exercises in SA,
surface attack bombing, and SAT, or surface attack tactics, learning how to use the F-16s' sophisticated defenses to avoid enemy
antiaircraft fire and SAMs while in final approach and targeting—
the two most vulnerable stages of a bombing run. Each course began with classroom instruction conducted by lead instructor Gary
Michaels and a specialist OTU, then actual flying, each IAF pilot
teamed with a personal instructor.

In the classroom the team once again studied the F-16's
weapons, communications, navigation, and mechanical systems,
from design specifications to operational capabilities. Since the
IAF pilots had already spent months studying the design, theory,
and mechanics behind the F-16s, on numerous occasions they
were much more informed about the technical aspects of the aircraft than even the instructors. Used to American "kids" just out of
college, the USAF instructors were surprised when the Israelis
peppered them with technical questions about the plane's design
and performance specifications. Oftentimes the OTUs had to refer
the pilots to General Dynamics engineers for answers. To their
younger American classmates, mostly in their twenties, the Israelis
carried a kind of cachet. Not only were they older, they had all
seen combat—with multiple kills. The United States had not been
involved in a full-fledged air war for a decade. In general only the
veteran instructors like Michaels, who had served in Vietnam, had
seen air-to-air combat. Even more than the IAF pilots' experience,
the U.S. instructors were impressed by their thoroughness and
their mastery of the aircrafts' mechanics.

But there were cultural clashes. The IAF culture emphasized pilot initiative and independent judgment. This created an independence and informality that could, on occasion, clash with the American military's rigid notion of order and protocol. Not long after the Israeli F-16s were delivered, Raz took one of the single-seaters out for a test run. During the flight he thought the navigation system was off. "Bumpy" was how he thought of it. He landed on the runway, taxied to the operational area, and parked the plane. He strode through the security checkpoint and returned to the squadron area to check the technical manual on the navigation system. He found the section he needed, put the book under his arm, and began the walk back to the aircraft.

Hill was laid out in two separate areas—the operational area, where the planes were manned, and the squadron area, where the flight personnel carried on the support business. The two areas were separated by a long cable some twenty inches off the ground. Personnel were prohibited from entering the operational area except through the security gate at the end of the field, about one hundred yards down the cable fence. Raz thought the extra hike down to the main gate a supreme waste of time, so he simply hopped the cable and cut directly across the field to his plane. In less than twenty seconds, two jeeps full of armed MPs screeched to a halt in front of him, their M-16 rifles pointed at his head.

"Hold it right there, mister," one of the MPs barked.

"But that's my plane . . ." Raz began.

"Hands up!" the MP cut him off, motioning upward with his rifle barrel.

"But—"

"Up!"

Raz raised his hands above his head and was arrested on the spot.

Meanwhile, Yaffe happened to be making his way from the

squadron area to operations when he turned a corner and, to his surprise, saw his squadron commander being escorted out of the area at gunpoint by three MPs. Proud and a perfectionist, Raz strode by, silently. Yaffe stared back in complete shock as the guards marched his leader away, his face set hard as granite, his eyes betraying a look of impotence, humiliation, and fury.

Several weeks later, after Gary Michaels and the base OTU officers had straightened things out between the Israelis and base security, Yaffe, Shafir, and Katz found themselves in another disagreement over base procedures. During a training run one of the planes malfunctioned. The Israelis reported the problem to the crew tech on duty in the operations hangar.

"We want to switch planes," Yaffe informed him. "Tell maintenance."

"We don't do that," the tech answered. "That's not how it works. You just don't take a new plane."

"Don't be ridiculous," Yaffe said. "We need a new aircraft."

The men argued with the tech until the crewman strode angrily out of the hangar. The Israelis mulled around the hangar for a minute or two until they heard the unmistakable sound of a jeep roaring to a stop outside. Seconds later three MPs marched to the pilots, the same MPs who just weeks before had arrested Raz.

"You again," Yaffe said.

The MPs were not amused. The IAF pilots argued their case, but the guards showed little interest in their troubles. They were breaking rules. That was all that was important. Unlike poor Raz, however, the men were not arrested.

At the end of training, the OTU officers threw Raz's team a going-away party. During the bash the lead instructor, Michaels, approached Yaffe, who had sent him scrambling on more than one occasion to find the answer to some complex design question.

Michaels smiled.

"You know," the instructor said, scratching and shaking his head. "It was a very *refreshing* experience having you here."

Raz's team returned to Ramat David in May to await delivery of the first four F-16s on Israeli soil, now scheduled for early July. As Raz, Yaffe, Katz, and Shafir set up housekeeping in the officers' quarters on the base, Amir Nachumi's team, including celebrated pilot Udi Ben-Amitay, left for their OT at Hill. At last the squadron received word that the F-16s would be delivered to Ramat David July 4. But the American pilots assigned to deliver the planes began complaining that they would miss Independence Day if they were held to the present schedule. To accommodate the USAF pilots, delivery was moved up to July 2. The American volunteer pilots flew the four F-16s in an eleven-hour, six-thousand-mile nonstop flight that required three in-air tanker refuelings. The IAF celebrated delivery of the planes like a national holiday, feting the American pilots as brothers. After more than a year, Israel's elite new fighting squadron finally had aircraft—a small but nonetheless crucial beginning.

Weeks after returning from the United States, Raz received orders from General Ivry to report to him at IAF headquarters in the Kirya in Tel Aviv. Raz flew in a small prop plane from Ramat David to the tiny air base at the northern edge of the city.

After some polite conversation about his time at Hill, Ivry shifted gears.

"I want you to begin training in low-level, long-distance navigation," Ivry said cryptically. "Start with shorter distances and work up."

"Yes, sir," Raz responded. "Anything specifically I'm looking for?" he added, fishing a bit.

"We want to know the extreme range of the plane's operational envelope," Ivry replied.

Raz said nothing, but Ivry had taken him by complete surprise. He had returned to Israel anticipating pulling together the new F-16 squadron and perhaps notching a few more Syrian MiGs with the sophisticated new fighter. The last thing on his mind was long-range navigation. What could Ivry be thinking?

The reports out of Baghdad were all bad. The nondescript, Soviet-era administration buildings that Khidhir Hamza had encountered on his first visit to al-Tuwaitha ten years earlier had disappeared completely, replaced by a vast, modern fortress more akin to Russia's Star City than anything found in the deserts of the Middle East. Officially dubbed Project Tammuz 17, the first of many such planned future complexes, the Nuclear Research Center covered one-quarter square mile and included dozens of labs, plants, shops, and buildings. In the north corner, landscaped with flower planters and hedges, stood the main administration building. Another building containing the main entrance hall, administrative offices, and a data center faced administration. To the south was the Italian-made fuel fabrication facility. Iraq's Atomic Energy claimed that the fabrication facility's purpose was to supply fuel for its ambitious power plant program, but Israel's scientists determined that the facility existed for one reason: to assemble uranium fuel packages for the production of weapons-grade plutonium. Just east of the fabrication plant was a radioactive waste facility to dispose of toxic detritus. In the southeast corner stood a large machine shop for manufacturing and repairing tools and hardware to support the reactors.

In the center of it all rose the mammoth thirty-foot-high dome of Osirak, officially renamed Tammuz I by Hussein. In his long-range plans, Osirak would be the first of numerous nuclear reactors built around the country. The Osirak reactor beneath the dome was essentially an open pool, thirty feet deep and filled with light water, which covered the plumbing, coolants, and the control and fuel rods. The light water would help modulate the rate of fission along with the control rods, slowing down the flow of free neutrons between the uranium pellets. The arching cupola functioned as an airtight vacuum to prevent radiation leakage. Scientists conducting experiments worked directly in the pool, manipulating the machinery by hand. Beneath the pool, ten feet underground, was the drive mechanism for the control rods and, extending west, the neutron guide hall, a large concrete room 60 feet long, 10 feet wide, and 30 feet high. Equipped with a twenty-ton bridge crane that traveled up and down the length of the room on steel beams, the hall was used to conduct separate experiments with free neutrons, which were siphoned from the fissioning uranium in the reactor.

A ground-level control room flanked the reactor pool and the dome. Within the building, adjacent to the reactor, was the Italian-made hot cell used to separate plutonium from uranium, which the Iraqis had renamed Project 30 July in honor of the Ba'thist revolution. Next door to Osirak stood the smaller French reactor, Isis.

A large guardhouse stood near the main gate, framed by a metal detector and X-ray machine. Soldiers in combat fatigues carrying AK-47s roamed the open spaces, which were ringed with closed-circuit television cameras. As Mossad had seen, the entire facility was fortified by a hundred-foot-high earthen revetment. Positioned at all four corners were AAA, antiaircraft armament, including batteries of Soviet-made ZSU 23mm guns on modified tanks, which fired four hundred rounds a minute. In between the AAA emplacements were Soviet-made SAM-6 surface-to-air missiles and radar-tracking units.

The photographs of the installation and reports out of Iraq had sparked a new urgency in Begin and the pro-raid ministers. For one thing, Iraq, exuding a new sense of invincibility, was no longer being as careful about hiding the ultimate use of its new nuclear program. In October 1980 the Iraqi daily *Al Thawara* ran an article about the Nuclear Research Center at al-Tuwaitha, reporting that Iraq intended the facility to be used "against the Zionist enemies." Meanwhile, Israeli intelligence estimates predicted that Iraq would have enough plutonium to produce two atomic bombs by 1982.

Pressed by Saguy and Hofi, Begin agreed to one final attempt at diplomacy. Israel's foreign minister Yitzhak Shamir called on the French embassy in Tel Aviv to warn the chargé d'affaires that Baghdad's French-built nuclear reactor could ignite a conflict in the region and set back recent gains in attaining peace in the Middle East. In a final plea, Begin sent a personal diplomatic letter to French president Giscard d'Estaing virtually begging him to pull out the French technicians from al-Tuwaitha and hold back from sending Iraq the remaining twelve kilos of enriched uranium. Giscard d'Estaing replied that he could not comply, but once again reassured Begin that France would never allow Iraq to develop weapons using the Osirak reactor.

A week later Mossad reported to Begin that Osirak would go hot within six months.

Begin made up his mind. About October 15 (the actual date remains classified), the prime minister called together a second secret meeting of the ranking cabinet ministers at his offices in Jerusalem. Yadin, Hofi, and Saguy continued to have serious objections to a raid. Yadin, in fact, had threatened to resign if Begin went through with the mission. Hofi and Yadin doubted that an attack could destroy the twelve kilos of enriched uranium France had already sent to al-Tuwaitha. Mossad had determined that the uranium was stored inside a concrete pyramid in an underground chamber located next to the neutron guide hall. Israel would be risking worldwide condemnation for nothing. Saguy believed that Israel still had no firm evidence that Iraq was yet capable of building an atomic bomb, and he was not persuaded that any perceived threat to Israel's security was justification for an unprovoked military attack on a sovereign nation.

"I do not believe fears of a 'Second Holocaust' justify the Israeli military taking any steps it thinks fit," Saguy told the assembled ministers.

Defense Minister Weizman had been so vehemently against a military raid that he had resigned in May in order to run against Begin within his own conservative Likud Party. Outside the administration, the Labor Party candidate for prime minister, Shimon Peres, and Labor Party chieftain Mordechai Gur had been leaked word of a proposed military strike against Osirak and were adamantly opposing any such operation, fearing it would endanger Israel's relations with the United States and the Europeans, isolating the tiny state.

Agriculture Secretary Ariel Sharon laughed derisively at that argument.

"If I have a choice of being popular and dead or unpopular and alive," Sharon scoffed, "I choose being *alive* and unpopular."

Since the first security cabinet meeting three years earlier, Begin had pledged he would not green-light a mission without the support of the entire cabinet, or at least the ranking political ministers. Hofi and Saguy were considered military, not political. The prime minister had grown impatient in the intervening years while Hussein continued to piece together his would-be atomic juggernaut. That Iraq could not produce an atomic bomb until 1982 or 1985 was beside the point to Begin. The important date was June 1981, when Israeli intelligence estimated Osirak would go hot, and after which Israel could not strike without the risk of causing widespread civilian casualties. Israeli scientists had estimated that destroying al-Tuwaitha and setting off a nuclear reaction could, depending upon the prevailing winds at the time, kill as many as one hundred thousand people as far away as Baghdad.

There was another pressing consideration as well: Israeli national elections were scheduled for the fall. Peres and Labor were enjoying a significant lead in the polls already. If Begin were to lose the prime ministry and a new government was formed, the opportunity to end Iraq's nuclear threat could be lost forever. The prime minister did not believe Labor had the stomach to deal with the crisis. In any event, by the time a new government was formed, the reactor would be up and running. Begin needed an endgame—and Ivry and Eitan, who had spent the last three years hammering out various plans and then discarding them, had finally delivered.

Dubbed Operation Hatakh Moshem, or "Ammunition Hill," after the famous '67 battle led by Doobi's uncle, the mission was to be carried out by IAF pilots flying F-16s, nonstop and without refueling, at low-level navigation from Israel to al-Tuwaitha. The mission had to be timed perfectly: the attack would commence at sunset on a Sunday to ensure the maximum safety of the French and Italian technicians who would be home on their day off (Israel miscalculated: the end of the workweek and the Iraqi "Sabbath," or day of rest, was Friday, and though many foreign technicians did take Sundays off, the plant was open for business). In addition, a late attack would give Israeli CSAR teams all night to rescue any downed pilots under cover of darkness. Of course, if the attack were launched *too* late, the ground would be too dark to distinguish from the horizon—a very dangerous environment for pilots on a bombing run.

Each fighter would carry two bombs. The target would be solely the Osirak reactor. How the F-16s would exceed their deadhead range of 540 to 560 miles to accomplish the 600-mile flight to Baghdad remained to be worked out, as would the type of ordnance used, the exact number of planes, the nature of tactical support, and myriad other details. But Ivry and Eitan assured the cabinet that the mission would be surgical and carry a low risk of casualties—at least, as low as could be expected in such a dangerous operation.

The unlooked-for addition of the F-16s, the thoroughness of the mission planning, and Ivry's assurances allayed many of the fears of the ministers who harbored doubts about a military operation. In addition, the political ministers, if not completely convinced of the wisdom of a raid, were loath not to support Begin, given the stakes. The prime minister made it clear that Saguy's earlier pronouncement that he would refuse to take any "responsibility" for a raid, even threatening at one point to withhold intelligence about Osirak from the IAF, would not be brooked.

After the hours of debate and squabbling Begin stood and looked down the table, his dark eyes flickering from the face of one cabinet member to the next. Some of these men he had known for four decades, had fought next to against the British in '47. He put both hands on the edge of the table and leaned in toward the generals and ministers, his chin up (some wondered later, was it *jutting?*), and announced, "There will be no other Holocaust in this century! Never. Never again!"

The ministers remained silent. No one dared oppose him—at least to his face. Ivry's mission was approved. He and Eitan were ordered to put the plan into action. No D day was set, but Begin made it clear he wanted it *soon.* November was set as a tentative date.

Raz and his squadron pilots at Ramat David continued long-range navigation training. Their numbers grew with the return of the second and third conversion teams from Hill. In the meantime General Ivry and his right-hand man, Col. Aviem Sella, one of the IAF's leading nuclear bombing and targeting experts, pulled together a secret ten-man operational team of engineers, scientists, computer experts, and combat strategists at IAF headquarters in Tel Aviv. Sella himself had served as an F-4 Phantom pilot at Tel Nof as a member of Israel's "black" squadron, a nuclear-weapons-capable wing assigned to the nation's ultimate defense. His experience in targeting and bombing was invaluable. As soon as Ivry had the green light, the operations team gathered in a top secret meeting that included Mossad and army intelligence analysts and their best nuclear scientists to pore over the blueprints of al-Tuwaitha obtained by Arbel's Paris station. The experts agreed that the key to the entire facility was the reactor itself—without it, the rest of the equipment was harmless. The actual target—ground zero—would be the reactor's thirty-foot-high dome, which was only several inches thick and composed of reinforced concrete.

Operations quickly discarded the idea of using so-called "smart bombs." These were mostly U.S.-made GBU-15s, which were dropped at a distance from the target and then guided by the pilot through remote-control movable fins and a television camera in the bomb nose. But the bombs were large and cumbersome, and the added weight and drag would reduce fuel efficiency, already a crucial factor in the long-range mission. Moreover, the new smart bombs were not 100 percent reliable and, worse, would demand the pilots' attention and add to their workload at the precise time they were most vulnerable to AAA fire. The success of the mission, Ivry was convinced, rested on simplicity and the element of surprise.

The team determined that the most efficient means of penetrating the dome and destroying the reactor beneath was to drop two-thousand-pound MK-84 slick, or "dumb," bombs, which simply used gravity. The bombs created a horizontal destruction pattern extending thirty-four hundred feet—more than enough to take out the entire reactor. They could be rigged for either instant detonation or delayed fusing and were simple, foolproof, and effective—something the pilots, taxed already with sophisticated mechanical and navigational systems, would welcome.

IAF wanted to be absolutely certain the bombs would do the job. No one had ever bombed a nuclear reactor before, and no one in Israel really knew for certain just what would happen if one were detonated. To find out, the IDF discreetly contacted two Israeli nuclear engineers at Haifa's prestigious Technion University, Joseph Kivity and Joseph Saltovitz, and recruited them for a mission. The scientists would travel to Washington and meet with representatives of the U.S. Nuclear Regulatory Commission, which licensed and regulated nuclear energy facilities in the U.S. The scientists could not, of course, tell the NRC officials the truth about what Israel really wanted to know: what was the most efficient means of destroying Iraq's unfinished nuclear reactor at al-Tuwaitha. Instead, Israeli intelligence created a cover story for the engineers. They would pose as representatives of the Israeli Electric Company, which supposedly was considering purchasing an electric-power reactor from the United States. The Israeli scientists would tell NRC the electric company was concerned about terrorism and threats from Arab neighbors and wanted to know what the effect would be if the facility were bombed.

The fact that Israel was deceiving its closest ally, indeed, its lifeline in the Middle East, did not seem to overly concern the planners. It was a defining distinction between Israel's political and military institutions and those of Western nations. The IDF, though highly trained and highly professional, nonetheless retained some of the seat-of-the-pants instincts, the risk-taking, that had marked the outlawed citizen army of the Haganah that had fought for the independence of the Jewish state. Such a serious mission, if considered by the United States military, would have occasioned months of bureaucratic second-guessing. Decisions in the Pentagon, as Vietnam had so vividly shown, tended many times to be driven more by fear of failure than will to succeed. Again, in the wake of the Iraq War, political considerations about how U.S. intentions might be perceived by Iraqis and the world community drove the military's occupying strategy rather than adherence to strict military tactics or even pragmatic solutions. The IDF, by comparison, seemed almost reckless.

Bypassing the usual diplomatic State Department or Defense channels, the Israeli embassy in Washington, D.C., contacted the NRC directly to ask for a meeting with the electric company scientists. A meeting was set up for October 9, 1980. The two engineers, Kivity and Saltovitz, flew to Washington on the eighth and met the following day with John O'Brien, James Costello, and Shou Hou in the NRC's local research offices. Kivity and Saltovitz wanted to know, specifically, what would happen to their reactor if, say, "a 1,000-kilogram [2,200-pound] charge penetrated [the] concrete barriers and detonated after penetration."

The officials, drawing from numerous studies conducted by the NRC, the U.S. military, and the Defense Department, detailed for Kivity and Saltovitz what systems in the reactor were most vulnerable to such an explosion and whose failure would result in "significant consequences," as Costello put it, and thus were "optimal targets for sabotage." In general, these were the reactor's fuel rods and the cooling systems. When Kivity and Saltovitz finished debriefing Costello, O'Brien, and Hou of all the knowledge and data they needed, the Israelis thanked their fellow scientists, shook hands, and then caught the first El Al flight back to Tel Aviv. There, they informed Ivry and his staff that the two-thousand-pound dumb bombs would be more than sufficient to destroy Osirak—if it were bombed before going hot. If the reactor were bombed while fueled with fissionable uranium, the NRC had confirmed, there would be danger of an uncontrolled reaction precipitating a nuclear event.

Back in Washington, after the NRC reps had time to think things over, the three men realized they had all noticed the same thing. As they put it in a follow-up memo drafted about the meeting: "Because of a lack of real interest in underground siting as a protective measure against sabotage, it was unclear whether the Israelis were interested in defending their own plants or destroying *someone else's.*"

Curiously, the NRC was not the only U.S. department suddenly having second thoughts about its dealings with Israel in the closing days of 1980. As Ronald Reagan became president-elect in November and William J. Casey prepared to take over the reins of the Central Intelligence Agency from Stansfield Turner, Langley (CIA headquarters in Langley, Virginia) began to hear disturbing rumors about Israel possibly compromising one of the nation's most jealously guarded assets.

To gather intelligence on Osirak, Mossad and IDF had been forced to rely on grainy ground-level photos secreted out of Iraq, the blueprints obtained by the Paris station, and HUMINT, human intelligence, gleaned by agents in Paris and Baghdad. Still, IAF had no comprehensive, big-picture surveillance of the entire facility or its environs. The service was forced to rely on old, sometimes outdated, maps and charts or, worse, on a subject's description of the area. Israel had no spy satellites orbiting the earth, snapping photographs of foreign bases and military installations hidden deep within the borders of its enemies. The nation had neither the budget nor the technology for such a sophisticated network.

But the United States did.

It was known as KH-11, the National Security Administration's supersecret, supersophisticated reconnaissance satellite. Launched December 19, 1976, KH-11 represented a stunning leap forward in technology—a sixty-four-foot-long satellite orbiting hundreds of miles above the earth, circling the globe every ninety-six minutes, relaying back high-resolution, digitally enhanced, real-time photographs so clear one could make out parked cars on the ground. As a reward, or more appropriately a carrot, for Menachem Begin's cooperation with Anwar Sadat at the Camp David summit, President Carter had granted Israel access to the internationally coveted KH-11 photographs in March 1979. Britain, which had been denied first-generation KH-11 intelligence because of a suspected leak in its communications intelligence establishment, was outraged that Israel was granted such access. The many U.S. defense and intelligence agencies that found they now had to vie with Israel for orbiting time were incensed as well. Israeli access would disrupt the delicate scheduling times that had been diplomatically hammered out between the various agencies over the years. Someone was going to be squeezed.

In Israel, access to KH-11 was seen as a monumental turnaround. For years during the Cold War, Mossad and the CIA had shared virtually all Middle East intelligence. So entwined were the agencies that in many respects the Israelis considered themselves virtual partners with the CIA. All that changed in 1977, when Stansfield Turner had severely curtailed the agency's liaison with Mossad. Convinced that the men in Carter's administration were naïve and anti-Semitic, Israel had responded in kind, cutting off its flow of intelligence about Africa and the Middle East.

Even now the Carter administration had put restrictions on Israel's KH-11 access. Israel could receive only I&W, intelligence and warning—that is, satellite photos depicting military activity such as troop movements or artillery placement occurring one hundred miles inside the borders of its Arab neighbors: Egypt,

Jordan, Syria, and Lebanon. It could not have regional surveillance of the entire Middle East. And certainly not Iraq. The idea was to provide Israel with defensive intelligence only. Any information that could be used to plan preemptive strikes was forbidden. So a routine had been established over the months. The military attaché at the Israeli embassy in Washington, D.C., would drive across the Potomac bridge to the Pentagon and, in an office under the direction of the Defense Intelligence Agency, pick up NSA's specially processed and carefully edited satellite photographs. These were then flown by diplomatic pouch to Tel Aviv where they were analyzed at the highest levels of Israeli intelligence.

From the beginning of the arrangement, longtime veterans of the American intelligence service anticipated that Israel would do everything it could to circumvent the restrictions. The Israelis would not disappoint them.

For starters, the agreement allowed Israel to make requests for special satellite intelligence. These would be handled on a case-by-case basis. Immediately the Israelis argued that the agreement did not pertain to common enemies of the U.S. and began pressuring NSA for full and unfettered access to all intelligence regarding the Soviet Union, including its supply lines into Iraq and Soviet training of Iraqi troops in western Iraq. The CIA turned those requests down. But Israel kept up the pressure. Mossad and the IDF had many friends deep within the agency. To these sympathetic ears they argued that Israel had to see all essential intelligence dealing with the Middle East, and only Israelis could know what was important to Israel. The Reagan administration had been a boon to the Israelis. The administration and CIA director Casey were far more sympathetic to Israel's arguments. To ensure that nothing was overlooked, Casey early in his tenure provided the Israeli liaison officer with a private office at Langley so Israel could have direct access to the intelligence officers processing real-time KH-11 imagery.

Over time, as the Israeli liaison established a friendly working relationship with the KH-11 officers, the original strictures became blurred. The Israeli officer was seen as an ally, relaying Israel's intelligence needs to the directors of the KH-11 program. What could it hurt to help them out on an informal basis? As a senior intelligence officer told author Seymour Hersh, "It was in our national interest to make sure in 1981 that the Israelis were going to survive." There was also the conviction that if Israel were refused intelligence, it would simply turn around and lobby supporters in Congress for the money to build its own satellite. By 1981, less than two years after Carter had first given them limited access to KH-11, Israel was extracting virtually any photograph they wanted, including satellite photographs of al-Tuwaitha. Israel had even managed to finagle a seat on the tasking committee to request its own flyovers. The mission pilots at Ramat David as well as most of the high command were never told of the existence of the photographs, but Mossad had seen them all. And so Hofi knew for certain that Osirak would be hot by midsummer 1981.

Soon after Kivity and Saltovitz returned from the NRC, General Ivry called Raz and Nachumi to Tel Aviv. The two pilots flew the sixty miles south in a fixed-wing prop plane, the farm fields of the Jezreel beneath giving way to the small villages and suburbs like Hod Ha'sharon and Ramat Gan outside Tel Aviv. Soon the winking warning lights showed atop the chimney standing sentinel at the north end of Ha' Yarkon Street. Four months had passed since the delivery of the squadron's first four F-16s to Ramat David. The squadron now had twelve F-16s of the Block 5 model, the original "Iranian" planes ordered in 1978. More planes from the current generation, Block 10, which included GD's new upgrades to the navigation, electrical, and weapons systems, were already coming off the production lines. IAF planned that ultimately there would be three F-16A/B squadrons. Raz would lead the first mission squadron of eight planes—though no one outside of high command was to know the target. The pilots had been told only to train in long-range, low-level navigation.

Raz and Nachumi took their seats in Ivry's office, and after the usual small talk, the general got quickly to the point.

"I want you to concentrate on an air-to-ground mission," he said, "air-to-ground" being militaryspeak for "bombing."

Nachumi saw Raz raise his eyebrows ever so slightly. Otherwise he showed no emotion.

"Remember, this is classified," Ivry said levelly. "You're not to discuss the mission with anyone. Not even your wives."

The men could have smiled. So far, there had been almost nothing to discuss. Following the first meeting in May when Ivry had told him to train for long-range navigation, Raz had gone straight to the Ramat David operations room and looked at the huge map of the Middle East mounted on the wall. Using a calculator, he quickly determined the optimum distance of the F-16 to be 560 miles. He cut a piece of string to match the map's legend of 600 miles, pinned one end of the string at Tel Aviv and then traced a 600-mile circle around the map. Looking at his rotating bull's-eye, Raz guessed somewhere inside Syria.

Ilan Ramon was the youngest of the eight pilots. He was twenty-seven and the only bachelor in the squadron. With thick dark hair and boyish good looks, he usually commanded the attention of any unattached female within hailing distance. But his good-natured, self-effacing charm and studious devotion to work also made him a favorite among the men. So did the fact that Ramon's mother and grandmother were both Holocaust survivors who had made their way to Israel after surviving the Auschwitz death camp. Ramon was acutely aware of his family's legacy, an unspoken yet profound conviction that seemed to bestow a kind of nobility to his youth and made his easy manner and openness all that more engaging. Ramon had been assigned as the squadron navigator. As a result, early on he had been apprised of the mission profile: a low-level, 600-mile flight with a tailwind to mission target; then a 600-mile, low-high return (which meant climbing and flying at high altitude) into headwinds. In the Middle East the prevailing winds are always easterly, blown in from the Mediterranean Sea, which sits off the western coast of Israel. Going with a tailwind meant flying east: looking 600 miles to the east on the map, Ramon was led to only one conclusion—Iraq. But where in Iraq, he wondered . . . Baghdad?

Hagai Katz, too, knew that a tailwind meant east, and that most likely indicated Iraq as the target. Then in September, Katz saw the headlines about the mysterious explosions at the Rome headquarters of SNIA, which was doing business with Hussein. Immediately he recognized Mossad's handiwork and quickly put two and two together to come up with the Osirak reactor at al-Tuwaitha. As the realization dawned on him, he felt surprised, apprehensive, and elated all at the same time. Though he felt sure he had divined the target, Katz, remembering his security agreement, vowed to keep it to himself.

Ivry meanwhile had anticipated all along that his pilots would be doing their own calculations, trying to guess the target. Though it seemed like overkill to worry so intensely about secrecy, surprise was the single most critical element of the raid—without it the chances of success were very slim. To throw the pilots off, during a briefing Ivry let slip a reference to "Habbaniyah," an Iraqi airfield west of Baghdad that was home base to Saddam's squadron of Soviet-made Tupolev fighters that had the range to reach Tel Aviv.

The controller's voice crackled in his headphones.

"1–3–3 Squadron cleared for takeoff."

"Roger," the pilot, a lieutenant colonel stationed out of Tel-Nof AFB in the north, replied.

Six F-15s had taxied out of their camouflaged hangars at the end of the runway and were holding on the tarmac, open to any intelligence satellites from the United States or the Soviet Union that might be orbiting overhead. The squadron leader barely gave it a thought. Like his superiors, he knew it was too late for any prying eyes to figure out what they were up to. The pilot returned to his preflight checkoff. The F-15 pilots did not have the luxury of the main attack squadron's F-16s and their computerized BITS, which automatically isolated and assessed each of the aircraft's weapons, navigation, electrical, mechanical, and communications systems. Instead, they had to manually check off each system one at a time, flipping switches and waiting for the green lights to blink back, a time-consuming process. As one of the support fighters that would shadow the attack group all the way to target, he was among the few who knew the mission destination. He had never flown into Iraqi territory. Not that it mattered.

His team was ready for takeoff now, an hour ahead of the mission squadron. They would circle west of Aqaba and wait for the eight F-16s to re-form, then follow them into Iraq. His job was to jam Iraqi radar over al-Tuwaitha with the F-15s' powerful ECMs and engage any MiGs that challenged the attack. Besides four Sidewinder air-to-air and four radar-guided Sparrow missiles, his plane carted 500 rounds of 20mm cannon fire that could be dispensed at the rate of 6,000 rounds a minute in short bursts. Enough to shred a MiG in seconds. At last the squadron leader gave the Go signal. He pulled back on his joystick, hearing the plane's twin engines rev and feeling the plane vibrate with the power as the first two hunter jets in front of him shot into the air and soared eastward. Following, he hit the afterburners and was pinned to his seat, his plane hurtling heavenward, for now blue, calm, and cloudless.

CHAPTER 4

THE WAITING

No plan, no matter how perfect,
survives first contact with the enemy.
—UNITED STATES ARMY MAXIM

For months, Operations' engineers and experts labored intensely over their computers in Tel Aviv trying to solve the mission's biggest obstacle—physics. Ivry and his staff had long ago rejected in-flight refueling as far too risky to attempt over enemy desert terrain. Besides, it had become a moot point because the Iranian-ordered F-16s had been designed to U.S. Navy specifications, which meant that the refueling baskets were on the bottom of the planes. Israeli tanker planes only refueled to the top of the aircraft, the same as the U.S. Air Force. The F-16s would have to get to Baghdad and back on one tank, so to speak.

General Dynamics test pilots, flying at fuel-conserving high altitudes and carrying no ordnance, had been able to extend the flight of the F-16s to two hours, fifteen minutes. The operational team, running numerous performance models based on planes flying low-level navigation and carrying two two-thousand-pound bombs one way, then flying high altitude with no ordnance on return, estimated flight time at three hours, ten minutes. The engineers and performance experts had to find another hour of flying time and fuel.

The physics of fuel consumption during flight are fairly basic: Discounting the vagaries of engine efficiency and pilot performance, fuel consumption is determined by two factors: weight and drag—that is, the physical resistance, or friction, to an object moving through air molecules. The more weight and bulk, the more drag and, thus, the more energy needed to propel the object.

For months the operational team worked up various performance drafts and modelings. With each new modeling, Raz's pilots would test the calculations in real time. The engineering team would then debrief the pilots, compare the real-time results with the computer modelings, and make the necessary adjustments. It was a long, stressful, sometimes dangerous process. And there were plenty of factors to account for. They tested fuel range with and without external wing tanks, with two or four A-As (air-to-air) missiles, and with four-plane, eight-plane, and twelve-plane formations. Flying in squadron formation increased fuel consumption because pilots were forced to maneuver in a group, varying their speed and vectors in order to maintain a constant distance between one another. Low-level flying consumed more fuel since air close to the ground is heavier than the thin air at high altitudes and requires more fuel to overcome the increased drag. Critical ECS, or electronic jamming systems, that allowed pilots to evade SAM radar and air-to-air missile tracking had to be jettisoned because the mechanism hung below the fuselage, creating more weight and drag. In addition, the jamming mechanism took the place of an external centerline fuel tank that could carry an extra 2,000 pounds of fuel, or 370 gallons. Something had to give—invariably it was pilot safety.

The long-distance flying posed practical problems for Israel as well. The country was only 210 miles long and 45 miles wide at its narrowest. That meant the pilots could not complete their long-range training flights within the nation's secure borders. Instead, Raz and his men had to begin their runs above the Mediterranean island of Cyprus and then follow the coast of Israel south, past the Gaza Strip and the Sinai to Sharm al-Sheikh at the southern tip of the peninsula, and then fly back the same way to Ramat David. The training flights to expand the long-range capabilities of the F-16s began at shorter distances and grew longer until matching the mission profile of 1,200 miles. Sometimes the flights were only an hour; other days they extended to three or four hours.

As far as the pilots were concerned, the majority of work was training their bodies. They were not prepared for the unexpected difficulties of long-distance flying. In the Mideast, enemy borders were dozens of miles away, not the hundreds or even thousands of miles distant they were for U.S. or British pilots. Damascus was only sixty-eight miles from Ramat David. In combat situations Israeli pilots had their weapons systems activated as soon as they were wheels-up. At the time the long-distance record was held by the IAF squadrons that had bombed the Suez Canal in '67. Most of Raz's pilots had never flown longer than an hour. Much of the initial training entailed building endurance, getting used to the bodily stress caused by extreme-range flying.

One of the pilots, Rani Falk, had just returned from Hill. An F-4 instructor at Hatzerim, Falk, along with Ilan Ramon and Relik Shafir, was among the youngest of the pilots. He had grown up in a farm village in the Jezreel Valley, not far from Raz's kibbutz, and like Zeev had spent days watching the planes soaring overhead to the nearby air force base. He was tall, broad-shouldered, open, and easygoing with a quick grin. He and Raz became close, perhaps because both men came from the same small village life and retained the same simple, straightforward values—honesty, hard work, and above all loyalty. Like Yaffe, Falk was a born pilot who had an instinctual ability to feel out his aircraft, to anticipate its response. But even for Falk, the long-distance flying could be brutal. He would return from his flights exhausted, grimy, clammy, and cranky, then have to attend debriefings to go over problems and mistakes. Sometimes the debriefing took longer than the mission.

Falk's first long-distance flight was a shock. When he reached Sharm al-Sheikh after what seemed an eternity, he was amazed to look up and see his INS (inertial navigation system) read just six hundred miles—only half the flight distance. After being cooped up in a cramped cockpit for hours—flying low-level, keeping eyes fixed on the wing leader, watching out for a sudden ridge or hill while continually checking instruments, maintaining distance and altitude, and flicking back and forth to the glass HUD as he also thought about fuel, risk, and the target—Falk's body felt as though he'd been beat up. Later the pilots learned that they burned so many calories, they lost from one to four pounds per flight.

Yaffe and Shafir quickly discovered another worrisome drawback. Flying in the 30-degree-tilted cockpit with knees pressed nearly to head level made trying to answer the call of nature impossible. For one thing, gravity worked against them. The urgency was made all the more pressing because the pilots were required to drink a great deal of water to stay hydrated. Since the fly-by-wire control stick was on the right side of the cockpit instead of in the middle, Shafir discovered that he was forced to try and open his pants zipper, already encumbered by a flight suit, with his left hand—a challenge he found impossible. Eventually he surrendered to the inevitable and urinated in his pants. As if that were not bad enough, the air-conditioning vent was exactly at crotch level, quickly chilling Shafir's wet lower parts to an excruciating degree.

At the end of one run, after he had touched down, Nachumi watched with alarm as Shafir came in fast at an unusually steep angle, hit the runway hard, threw on the afterburners, and jammed on the brakes, trailing a blue cloud of smoking rubber. He popped the cockpit, jumped down, and ran to the side of the runway. Nachumi, thinking the plane was on fire, radioed for an emergency vehicle and quickly climbed out of his plane. Bounding down the runway, he arrived to find Shafir stooped down in the weeds, surrounded by the emergency crew.

"Sorry," Shafir said, looking embarrassed but relieved. "I had stomach problems."

Hagai Katz, ever thorough and organized, took the time to assemble a homemade instruction manual for the pilots. Included were checklists of everything the pilots might need during flight— how to adjust the weapons systems, the radar functions, the navigation instruments. He made photocopies of the lists, then inserted each list into its own plastic sleeve and gathered them all in a notebook, like a photo album. He gave one of these notebooks to each pilot to use as a quick reference guide in flight.

One of the pilots was Amos Yadlin, an F-4 major who had recently returned from Hill with Falk and the third group. Tall, thin, with a full head of Kennedy-like brown hair, Yadlin appeared almost professorial, a look that complemented his quick, perceptive mind. He was also a seasoned combat veteran, seeing plenty of action in the Yom Kippur War. In some ways Yadlin was an easiergoing version of Raz. But Yadlin had a devilish side as well. One day when the teams were returning from another trek down to Sharm al-Sheikh, he jogged up to Katz standing on the tarmac, yelling, "You saved my life! Your checklist saved my life."

"How?" Katz asked, gratified and, to be honest, a little surprised his notebook had come in handy so quickly. "I just made it."

"Well, I had to pee," Yadlin replied. "I couldn't think of what to do. And then I remembered your checklists. I opened up one sleeve at a time and peed into them like a cup."

Yadlin grinned as Katz looked at him in horror.

"They held the entire load!" Yadlin added proudly.

The squadron pilots overhearing the conversation broke into howls of laughter. Chagrined, Katz marched back down the tarmac to the briefing room alone.

Months into training it had become obvious the F-16s would have to carry two external fuel tanks, one under each wing. Designed by GD, each tank added an additional 3,000 pounds, or 450 gallons, of fuel. But carrying the two detachable tanks and two 2,000-pound slick bombs, the fighters had room for only two Sidewinders, one at the end of each wing, instead of the usual four. And, as already decided, there would be no jamming devices. The pilots would be at an even worse disadvantage.

The fuel problem was still not solved, however. During training flights Raz's pilots were consistently running short of fuel, even with the wing tanks. The only thing that could help was to carry an optional centerline fuel tank, which held another 2,000 pounds, or 300 gallons, of fuel. But there was a problem: Israel had no centerline tanks. The U.S. Defense Department had excluded centerline tanks from the trade agreement. Since the planes had been sold to Israel on the strict condition that they be used for defensive purposes only, there was no reason, in the opinion of the U.S. Defense Department, that the IAF would need such tanks for long-range flying. Ivry immediately put in a plea for twelve centerline tanks. The Israeli Defense Ministry, the Foreign Ministry, and the ambassador to the U.S. all went to work intensely lobbying the U.S. to sell Israel the centerline tanks. In the meantime all Ivry could do was wait . . . and sweat.

The northern commander at Tel-Nof Air Force Base near Galilee, Gen. Avihu Ben-Nun, saw a new opening for his F-15s. The IAF had finally convinced the United States to sell Israel the F-15 conformal fuel tanks. Fastened to the fuselage at the base of the wings, the tanks would give the F-15s the range to reach

Baghdad and back. Once again Ivry was forced to fend off another challenge to the F-16s as Ben-Nun argued to Eitan and high command that his F-15 squadron should be given the mission. Ivry countered that they had already progressed far into mission training at Ramat David. Ben Nun took Ivry's refusal to make a switch as a personal rebuff. Meanwhile, word leaked down the chain of command, and Raz's squadron grew anxious that the mission—whatever it was—was going to be pulled out from under them.

And then things got complicated.

Ever since Ayatollah Ruhollah Khomeini had landed in Tehran in February 1979, Saddam had kept a jaundiced eye to his east. Socialist and secular, Hussein distrusted the bearded, fanatically religious Shi'ite Muslims who ruled Iran and made up the majority of the population in the southern half of his own country.

"This place hardly seems like part of Iraq," Khidhir Hamza recalled Hussein grousing one day as a mob of Iraqi Shi'ite demonstrators chanted Khomeini's name in the streets. "They don't even speak Arabic."

On September 17, 1980, Hussein, convinced that Iran was plotting his assassination with Iraqi Shi'ites, canceled a 1975 peace treaty with Iran and invaded the disputed Shatt al Arab estuary in the north of the Persian Gulf that formed the border between the two countries. Hostilities quickly escalated, and by September 22 the nations were in a state of all-out war, conducting air and large-scale ground assaults.

The evening of September 30, Ivry was still at work at Tel Aviv headquarters when he was informed that at least two Iranian F-4 Phantoms had just bombed al-Tuwaitha. Intelligence was still trying to get details, but initial reports indicated that the bombs had missed the reactor and damaged some laboratories and support facilities. The most serious blow was to Osirak's water-cooling system and plumbing, which took a direct hit. In the end the damage was minor. Begin was furious, cursing the incompetent Iranians who could not "finish the job." Ivry was also disturbed, but for a more practical reason. In response, Iraq put all its antiaircraft defenses on "alert time," meaning the readiness time of al-Tuwaitha's AAA batteries was significantly heightened. And, as an extra protection, Iraq launched a ring of tethered balloons twenty feet high around the Nuclear Center's walls to interfere with low-flying bombers. Ivry's already impossible attack plan was, if possible, now even more difficult. What could go wrong next? he wondered.

Fearing more strikes, the following week France and Italy ordered the two hundred techs and engineers employed at al-Tuwaitha to evacuate immediately. Mossad reported that the workers, who lived with their families at a separate compound away from the center, were packing up and heading for home. By November, Mossad was reporting that work at Osirak had come to a complete halt. Khidhir Hamza and his administrative colleagues continued to come to work at al-Tuwaitha, but the constant activity around the reactor, the buzz and comings and goings of the construction crews and the nuclear techs, had all but disappeared.

With the immediate threat of enriched uranium production over—at least for the time being—Begin, under pressure by Saguy and Hofi, called off the mission.

Raz and the F-16 pilots continued training, however. None of the pilots knew what the mission was, let alone Begin's decision to postpone it. But details were beginning to be revealed. For the first time, Raz informed the pilots of the kind of ordnance they would be carrying: two 2,000-pound dumb bombs. Because of the sensitive placement of the target, he told them, carpet-bombing was out.

"It will be a visual drop. You will need perfect accuracy," Raz

said. "The target is heavily defended by multiple AAA emplacements and SAM-6s."

The men would have to perform pinpoint targeting while avoiding withering AAA fire.

Flying over the desolate Negev, Raz's squadron practiced individually at first, diving at between 35 and 40 degrees. The pilots
used BDUs, thirty-three-pound dummy bombs that exploded with
white phosphorous smoke so pilots and ground personnel could
mark the accuracy of the drops. For targets the IAF used painted
circles on the ground and, later, old Sherman tanks. To approxi-
mate the huge dome of Osirak, Ivry's command later had the F-16s
practice diving at a huge, secret IAF radar dome located in the
Negev, though the pilots were not told why they were practicing
bombing an Israeli radar dish. The squadron also made several
flights dropping live MK-84s on desert targets so they could experience the shock waves and the extent of the frag pattern.

Targeting demanded absolute concentration. The pilot had to
fly dangerously low to the ground, constantly looking for unmapped peaks and outcroppings, or even telephone wires, next
pop up to ten thousand feet, nearing the speed of sound, and then
dive on the target, switching on the weapons system, all the while
checking the overhead HUD and being careful to line up the
bombsight with the target. After release, turning radically, the pilot
would blast off into the ether like a bat out of hell, breaking the
sound barrier and streaking to the safety of high altitude, praying
that a SAM was not behind him, trailing the heat of the afterburners to soar literally straight up his tailpipe.

Falk thought of it as the "moment of truth." The flying, the
flesh-flattening Gs of right-angle turns, the diving, evading—all
the air acrobatics—came to him naturally, as smoothly and easily
as an opening aria came in the silence to the mind of a Mozart.
But bombing was something else. It was the payoff, the entire

point of the mission. To miss, to fail in front of your fellow pilots, your peers, was devastating. You failed yourself, your team.

Early on in training, Falk missed a target during a practice run. He felt so bad, he did not even want to land. He wished he could keep on flying . . . just disappear. Instead, he had to land, trudge to the briefing room, and explain why he had missed.

To make the training as close to real time as possible, the mission team conducted combat games, with the F-16 squadron the Blue Team and a wing of F-15s, standing in for Iraqi MiGs, the Red Team. During the bombing runs the Red Team would try to intercept members of the Blue Team, forcing them to evade and then begin targeting. In real life, over Osirak, the pilots would not have enough fuel to engage in a dogfight and then expect to make the return trip home. A quick evasion was their only hope of completing the mission and returning to base.

While the pilots practiced targeting, the operational team worked out the details of the bombing run. Precision bombing, or pinpoint targeting, was a fairly sophisticated technical undertaking, necessitating exact mathematical calculations and modelings. The two crucial elements were the IP, or initial point, and pop-up. The IP was the exact agreed-upon location, usually some three or four miles from the target, at which the aircraft would begin its climb. The climb was called pop-up. Bombing from a flat approach was out of the question; a bomb released at almost the horizon line would ricochet off the concrete dome. Instead, the pilots would pop up—that is, pull the nose of the plane up, hit the afterburners, and climb to an altitude of between 8,000 and 10,000 feet in order to begin an angled dive at about 30 degrees.

The bombing run consisted essentially of seven elements, or timings. After pop-up, the second element was the pull-down altitude, the preset altitude at which the pilot pushed the nose down and began the dive toward the target. It was imperative that this elevation be as low as possible without endangering the accuracy of the dive because a shorter dive distance lessened the pilot's exposure to AAA and SAM fire. The third point was the apex altitude. This was the altitude the plane climbed to during the fraction of a second it took the electrical impulse from the control stick to reach the plane's mechanics controlling the wing flaps. At six hundred miles an hour, the F-16 could cover a considerable distance in a split second. The apex altitude was the exact point at which the dive would start and, like the pull-down point, it had to be as low as possible. The fourth element of the bombing run, called tracking on final, was the actual dive itself, measured from the apex altitude to the release point, the fifth element of the bomb run and the altitude at which the ordnance was dropped. The pilot kept his bombsight lined up on the target until the pipper, or "death dot," covered the target completely. At that moment he squeezed the red button on the control stick and the bombs were released, or "pickled," off the wings. The time of tracking on final, from high to low as pilots referred to it, had to be between three and five seconds.

As soon as the pilot released ordnance, he initiated the sixth element, recovery, or escape, firing afterburners and trailing thruster flames, climbing to high altitude. As the pilot began his escape and climb, there was another split second between the time he pulled back on the fly-by-wire control and the actual response of the aircraft mechanics. During this fraction of a second the plane would "settle," or sink, to a lower elevation. Predetermining this lowest point of the dive, called the recovery altitude, was absolutely essential, since the pilot had to avoid the frag pattern, the bloom of shrapnel and debris following detonation, which, in the case of the MK-84, rose 2,400 feet in nine seconds. The pilot, if not careful, could easily blow up himself and any fellow pilots following too closely. The final element of the bombing run was the escape maneuver, during which the pilot could hit a body-crushing eight Gs while negotiating radical 90-degree turns and climbing to 30,000 feet to defeat SAMs.

To compute the precise distance from the target to initiate pop-up, what angle to start tracking on final, the exact altitude of apex, and all the rest, Avi Sella's team thumbed through engineering books thick as IRS tax rolls, poring over computer graphs and charts and physics tables to check and recheck their figures. The concept was basically computing backward: first determine the altitude of the frag envelope, then add the recovery time, the pull back, then tracking on final, and so on. Once these figures were computed and added together, the sum determined the exact altitude to the meter at which the squadron commander would begin his dive. Each pilot would then, in turn, follow precisely.

Operations initially determined that pull down should start at around 8,000 feet. The U.S. Air Force routinely added a cushion of 500 feet when computing their recovery altitude. But the Israelis, fearful about the heavy AAA defenses, were determined to squeeze every fraction of a second out of the tracking time. Sella's team figured that because the bombs would pierce the reactor cupola and fall through before exploding inside, the dome itself would function as a shield, cutting the frag pattern in half. The decision was made to press the attack to the absolute minimum distance, with no safety net. To add a degree of security and ensure that follow-on pilots would continue to have an unobstructed view of the target, the bombs of lead pilots Raz and Yadlin would be rigged with delayed fuses.

Behind every maneuver was the element of speed, sometimes suicidal speed. After weeks of test dives and modeling and more test dives, Raz and his team were able to cut the apex altitude to only 5,000 feet, shaving off valuable seconds in both pop-up and tracking. Every second bought them more time to drop their bombs before Iraqi AAA gunners could fire up their radars and get their weapons systems operational. One day in late December, Raz briefed the team that they would be targeting white bull's-eye circles on the ground, diving in five teams of two. Shafir was paired with Yadlin. The attack profile called for Shafir to follow Yadlin at a one-second interval—or, at 360 knots, about 200 meters. The pilots fired up their engines, ran through the computerized checkoff, then lifted from the Ramat David tarmac and soared southward toward the Negev. Yadlin popped up and began his dive with Shafir just behind him and to the side. But as Yadlin reached recovery and began pulling out, Shafir was too close and the two planes headed for collision. Shafir careened radically to port, the G forces pinning him against his seat. The two F-16s streaked by each other, the air crackling in thunder. Once back at the base, Raz and Sella's team reran the computations and quickly adjusted the profile, extending the follow-on and adding a margin of safety.

What was the point of avoiding AAA if all the pilots accomplished was taking one another out?

In January 1981, in a somewhat unusual move, Ivry visited Raz and Nachumi at Ramat David.

"I know you have wondered where you are going," Ivry said. "Now I will tell you: Your target is the Osirak nuclear reactor at al-Tuwaitha near Baghdad."

Neither pilot said a word. Both acted as though they had known it all along. But inside, Raz felt his stomach flip. He was sure no one had ever bombed a nuclear reactor before. He knew immediately that the mission would be historic, something schoolchildren, their children, might read about someday. What were the chances of success? Pretty good, he decided. They had trained for this for six months already. It was just a job. Before he left the briefing room, Raz had already filed it into his mental box and put it away. Worrying was not going to change anything.

Several weeks later, on a gray winter afternoon, Yadlin's wife, Karen, spotted two grim-faced IAF colonels pull to the curb in front of her home and start up the sidewalk. Like all pilots' wives, it was the moment she lived in constant dread of. She knew their appearance could mean only one thing: Amos had been killed in the line of duty. She was wrong. The messengers of death veered up the sidewalk and knocked on her neighbor's door, the home of Udi Ben-Amitay, a member of the second team and one of the initial twelve F-16 pilots. He was also one of her husband's closest friends. Yadlin and his wife would get together often for dinner or Sabbath with the Amitays. Udi had been scheduled to participate in a training dogfight. In this instance, the Red Team, the enemy, comprised two F-4 Phantoms opposing the Blue Team, made up of the leader Ben-Amitay and a second F-16. Tragically, the F-4 leader and Amitay maneuvered too close and collided in midair. Both men were killed instantly. Amitay was the new squadron's first fatality. There would be an empty desk in the briefing room. Yadlin and the rest of the squadron were devastated when Colonel Spector broke the news to them. The wives, who shared a world much closer than neighbors, gathered to comfort Amitay's wife, taking turns cooking meals and watching the kids. The entire squadron attended Amitay's funeral. His death had created a hole in the unit, but Israel was losing many pilots at the time, both to war and to training accidents, as the IAF struggled to find tactics to overcome the technical advantage of Syria's new SAMs.

Commander Spector was both a symbol and a source of strength for many of the pilots at Ramat David. In a nation of military heroes, Iftach Spector was renowned above all in the IAF. At just forty-one, he had chalked up more combat kills than any pilot in history, having served in the '67 war, the '73 war, and the War of Attrition. During the Yom Kippur War alone, Spector had single-handedly shot down fifteen MiGs—an almost incomprehensible number in a profession in which veteran pilots the world over sported medals for shooting down maybe two or three enemy planes in a lifetime.

He was also no stranger to controversy. On the fourth day of the 1967 war, Israel dispatched a squadron of Mirage fighters off the shores of Gaza, allegedly to confirm reports of an Egyptian gun-boat. Instead, the Mirages spotted the USS *Liberty,* a high-tech, audio-surveillance ship some miles off the coast. Despite the fact that the *Liberty* was clearly flying U.S. colors and had a score of U.S. Navy sailors on its topdeck, the Israeli squadron leader iden-tified the ship as a "hunt class destroyer" with "no markings" and ordered an attack. The Israelis strafed the cruiser three times, killing eight men and wounding twenty, including the ship's cap-tain, shot in both legs. In the storm of protest that followed, Israel apologized profusely, insisting it was a mistake. Ultimately the gov-ernment paid $12 million to the victim's families. Many in the Pentagon, however, remained unconvinced. They suspected Israel did not want the United States picking up information about its operations in the Suez. Relations between the two militaries re-mained at an all-time low for years. The Mirage commander who had led the attack on the *Liberty* was none other than Iftach Spector.

Surprisingly soft-spoken but with quick, penetrating dark eyes, he was a collection of contradictions. Named "Iftach" after the tragic judge of the Old Testament, the misbegotten son of a harlot forced to sacrifice his daughter as the price for a desperate victory to save the Hebrews, Spector seemed to have inherited the tragic smile and sad eyes of the doomed prelate along with his name. As though acknowledging as much, he would occasionally turn his sad eyes mischievously on his interlocutor, a hint of an ironic grin crinkling at the corners, and pronounce innocently, "I don't know why they called me that." At the same time he was well aware of his near-mythic standing in the air force. He carefully nurtured the image, carrying himself with regal bearing. The effect made him a figure of strength but also of openness to the men who served under him. Indeed, to Nachumi, Katz, Yadlin, and the younger pilots, Spector was like a god. He could do no wrong.

As base commander, Spector was responsible for everything on the air force base, not only the planes, personnel, pilots, and squadron commands but niggling nuts-and-bolts details from mess supplies to infrastructure maintenance, all the while remaining an active fighter pilot and squadron commander involved in tactics and mission planning. When the new F-16 Fighting Falcon squadron was constituted at Ramat David, Spector became commander of that unit as well. It was an all-consuming responsibility.

Nevertheless, Spector made himself the squadron's first pupil. As the commanding officer, he felt it a weakness to have men under his command who were expert in areas he knew nothing about—especially men he had mentored. As base commander he was one of the few allowed to know about the secret Osirak mission. A week after the first four F-16s were delivered to Ramat David, he began training in the Falcon. He worked up his own solo modelings, practiced simulated attacks, flew low-level, long-range navigations south along the Mediterranean coast and down the Sinai Peninsula. He said nothing to the men about this. But Raz was a little annoyed. What was Spector doing flying the F-16?

As Raz suspected, Spector had a plan. He knew the historic importance of the mission. A man used to the spotlight, he knew the notoriety it would bring. How could he, the nation's most renowned fighter pilot, the commander of these men, stay behind while they flew into certain danger—and perhaps immortality? It would look like he was shirking.

"I want to join the Osirak mission," Spector told Ivry in his office.

Ivry was stunned.

"I am their commander," Spector said. "It is my duty to take my place with the men in this mission."

"But you have not had the conversion training of the other men in the F-16," Ivry replied.

"I have trained myself. I am as ready as anyone," he argued. "I command many missions. Why would I not be part of this? It would be inappropriate for me, as their commanding officer, to remain at base while the men under me risk their lives on this mission."

Ivry had long respected Spector as one of the IAF's greatest pilots. But his gut told him Spector was not prepared. The mission pilots had all been carefully selected by him personally. They had trained hard for nearly a year, had logged hundreds of hours, had come together as a team. Spector would be in over his head in the new plane. And adding Spector to the mission would mean that one of the other pilots would be shoved out, pulled at the last second because an air force bigwig suddenly decided he wanted a piece of the action. It wasn't fair, and it wasn't good generaling.

"No," Ivry said. "I'm sorry. I have to deny your request."

Spector was stunned. He couldn't believe it. He flew back to Ramat David chewing over what he should do. Could he just walk away? How would it look? No, he thought he had to be on the team leading his men. He decided he would have to go over Ivry's head and make his request directly to the chief of staff, Raful Eitan. It was a fateful decision that would ultimately affect many people—but no one more than Spector.

Spector made an impassioned plea to Eitan to intervene. The

chief of staff was in an impossible position. He had known Spector for twenty years and respected him immensely, both as a pilot and friend. How could he possibly humiliate him, Israel's most celebrated fighter pilot? On the other hand, Eitan held Ivry in no less regard. And how could he undermine a commanding general by overturning his decision, completely violating the Israeli military's sacrosanct chain of command? More critical, inserting Spector into the mission at such a late date could be dangerously disruptive. It was a lose-lose proposition however he cut it.

From the beginning the F-16 squadron had been under the command of Zeev Raz. As originally envisioned, the Ramat David F-16 wing would initially be broken into two squadrons, the 117 under Raz and the 110 under Nachumi, as more F-16s arrived from the United States and more pilots were trained. The mission team, on the other hand, consisted of eight pilots and two backup pilots, made up from the first three conversion teams sent to Hill. After Ben-Amitay's death, the initial squadron consisted of Raz, Amos Yadlin, Doobi Yaffe, Hagai Katz, Amir Nachumi, Relik Shafir, Ilan Ramon, and Rani Falk. As within any group of very competitive men, there had been some jockeying for position and leadership. By late 1980, Nachumi had been chosen by Ivry to head the second F-16 squadron, the 110.

Since first conceived, the mission profile had gone through several modelings. At one point, when Begin pressed for an early mission in November 1980, Operations thought they could go with only four aircraft. When word of this came down the line, Nachumi grew agitated, fearing that he and his group would be bumped off the mission. He lobbied Spector hard, then began flying regularly to Tel Aviv and haranguing Ivry and IAF command to include his team in the raid. Day after day Nachumi appeared at Kirya, arguing his case, asking if they had made a decision yet.

Finally, Ivry, annoyed, snapped at him. "Okay, okay, you're in. Now go away!"

Nachumi pointed out that early on, IAF had planned that the mission would be made up of two formations of four planes, one drawn from each squadron, Raz's 117 and Nachumi's 110. Nachumi argued that missions were assigned to squadrons, not men, so he and Raz were equal. Raz—and the majority of the pilots—considered himself the mission leader and Nachumi the leader of the second team. This rivalry caused some friction between the pilots. Some on Raz's team considered themselves one eight-man attack squadron working together. Why split it up? The men found themselves constantly gravitating between Nachumi and Raz. Both were terrific pilots and both had very strong egos. But there was a distinct personality difference between the two. Each had followed a different career path to arrive at command. Nachumi came from Spector's crack Phantom group in Beersheba in the south. Raz had moved up through northern command fighting in Syria and the Golan Heights and was close to Ivry. Raz was detail-oriented, no-funny-business, acutely sensitive to any challenge to his authority. Nachumi was more outgoing and not at all reticent about his accomplishments and his talents. This unspoken rivalry brought a sharper edge to the inevitable competition between two strong, ambitious leaders and, in turn, their two competing squadrons. It fell to Iftach Spector to play the diplomat, smoothing things over, controlling a rivalry that could actually be healthy by keeping the men focused and finely tuned.

Spector's move to force himself into the mission, and going over Ivry's head to do it, tipped this delicate balance and brought some of those vague, lingering feelings below boiling to the surface. Ivry, understandably, was annoyed. He could not believe that Spector would have the "bad form" to go over his head. It was disrespectful. Raz and Falk were furious. It was likely that Falk, who had rotated into a secured slot when Amitay had died, would be bumped to backup status if Spector were assigned to the mission. He couldn't help resenting it. The men had spent nearly a year training for the mission together. Now at the last minute the base commander wanted to walk in and grab a spot. Falk thought to himself, "Hey, come on, you had your time. Give it to the kids."

As for Raz, he had never held Spector in any particular awe. If anything, Raz was probably a little contemptuous of Nachumi and the others' reverence for the commander. He respected Spector's combat record and career as much as the next man, but who was he to think he had the right to simply walk onto Raz's squadron? He never even approached Raz, the group leader. To add insult to injury, Spector would knock his friend Rani out of the mission.

Even the men who had more or less been Spector's disciples— Yaffe, Katz, and Yadlin—felt that adding the commander to the mission at such a late date was not a good idea. Despite his peerless abilities as a combat pilot, the fact was, Spector did not have the expertise in the sophisticated F-16 that the other pilots did. And it was not fair to bump one of the men who had trained so long and so hard. As a group, the pilots met with General Ivry and informed him of their opinion.

Despite the opposition, General Eitan could not bring himself to disappoint his most heralded commander. He called Ivry and, as much friend as superior, asked the IAF head to make room in the mission for Spector. After all, he was already the men's commander. Ivry relented. But, he insisted, Raz stayed mission leader.

Ivry informed the squadron leaders in person. Iftach Spector was a member of the Osirak mission.

"I won't have him in my squad," Raz bridled.

An awkward silence fell among the airmen.

"He can be my wingman," Nachumi said.

Spector was assigned as second-in-command of the second team and as sixth pilot in the bombing run, following second team leader Nachumi. As the senior officer, it was a bit awkward and certainly unconventional for the commander to be under the command of junior officers, but Spector was not going to complain.

As simple as that: Spector was in, Falk was out. Falk was respectful, but decidedly cool around the commander. Raz, all business, swallowed the decision and moved on. But, it would turn out, the squadron was transformed in a much deeper sense than anyone suspected. For the first time, it became obvious that there were now *two* teams. And two team leaders. Spector was no longer just the base commander: he was a member of the second group, Nachumi's team. The commander was aware of the latent animosity, but he was determined to overcome it. He took his place on the mission team. He was friendly, unassuming. He respected each of the pilots and was sure, given time, that they would accept him. In any event, he wasn't going anywhere.

Was it a matter of ego or, as Spector told himself, a conviction that a leader's place was in the line of fire with his men? Or was it perhaps something less conscious—ego, an inability to let go, to miss the spotlight? Not even Spector knew the answer to that. Whatever the case, it was too late to turn back.

By March 1981, Israel received word that the U.S. Department of Defense had agreed to sell the IAF twelve F-16 centerline fuel tanks. Operations recalculated all the data and factored in the extra fuel accorded by the centerline tanks. When the engineers were done crunching the numbers, the news was not good. At an average airspeed of 331 knots, flying fifty meters above the ground, given the prevailing temperatures, humidity, and wind patterns of the route and taking into account all the extra weight of the fuel itself, including the two external wing tanks and the centerline tank, Operations calculated that the Pratt & Whitney single engines would burn 4,940 pounds of fuel an hour. Factoring in the radical fuel intake when the afterburners were used during takeoff, pop-up, and escape, the operational engineers estimated that by the time the pilots reached the Euphrates River, their aircraft would have already burned through 9,000 pounds of fuel. That left only 6,000 pounds of fuel to get home on—if there were no intercepts or evasions. Indeed, during test flights the pilots were coming up short some forty to sixty miles.

Somehow they had to find another sixty miles—this after already virtually stripping the planes clean. The support team checked and rechecked their modelings, scanned the performance specs. In the end they came up with two last-ditch ideas—both risky. The first was to jettison the wing-mounted fuel pans over the desert as soon as they were empty. That would lighten the planes by several hundred pounds, cut down the drag caused by the hanging tanks, and save as much as ten minutes' flying time. Indeed, General Dynamics had designed the external fuel pans, which looked a lot like bombs themselves, to be released from inside the cockpit. But there was still a danger. The fuel pans hung beneath the wings next to the two-thousand-pound bombs. The tanks and their wing clips were not designed to be released while the aircraft was carrying ordnance. There was a real risk that the pans, let loose at three-hundred-plus knots an hour, could easily collide with the bombs, damaging their release clips or, worse, causing the bombs to detonate. The pans could also be caught in the updraft and flip up and over the wings, causing damage to the wing flaps.

As weapons officer, Katz was particularly concerned about the idea of jettisoning the external tanks so close to the ordnance. He called the chief design engineer at General Dynamics in San Diego and asked him what he thought the chances were for dumping the fuel pans in flight while fully armed. The engineer rechecked the design specifications and told Katz he thought they could get away with dropping the tanks if they kept their airspeed under four hundred knots. The issue was settled: the wing pans would be dumped over the Saudi desert.

The second idea was to do a "hot refueling" on the runway at

Etzion. With the engines running, spewing hot streams of jet exhaust, the F-16s' tanks would be topped off on the runway by fuel trucks before takeoff, replacing the hundreds of gallons of jet fuel burned while conducting checkoffs and taxiing. It was a dangerous procedure, with a risk of the hot exhaust igniting the fuel and exploding the tanker trucks or the F-16s. Once again the book said it could not be done. They would do it anyway.

By the end of March, Mossad reported to Begin that the foreign workers were returning to al-Tuwaitha and building had resumed at Osirak. France and Italy decided the Iran-Iraq War was likely to drag on for years, bogged down on the border in World War I–style trench warfare. The likelihood of another Iranian air strike on al-Tuwaitha was minimal.

Begin wanted the air strike back on and began lobbying the ministers for final backing. Deputy Prime Minister Yigael Yadin was still very much on the fence, even though he had not challenged Begin outright at the October meeting. Taking no chances this time, Begin did behind-the-scenes arm-twisting. Yadin had been a member of the '74 blue-ribbon panel chosen to investigate the intelligence failings that had allowed the Israeli military to be taken by surprise in the opening days of the Yom Kippur War. At that time he had been supplied with intelligence material that had been skewed and doctored. Now, for those reasons, Yadin did not trust the military and Mossad intelligence estimates of Osirak. He insisted on seeing the raw data, the original classified reports from the field agents. In early March, Begin arranged for Chief of Staff Eitan to meet secretly with Yadin and show him the raw data. At the meeting, Eitan presented Yadin with the classified Mossad reports and photographs, including the top-secret KH-11 satellite shots that clearly documented the return of the foreign techs to al-

Tuwaitha and the resumption of the construction work. By the end of the meeting Yadin agreed to withdraw his opposition to the raid.

The prime minister called for a top-secret security meeting on March 15, 1981, to be attended by all ten ministers. Several of the ministers, including Health Minister Eliezer Shostak and Deputy Defense Minister Mordechai Tzitori, remained deeply uneasy about an attack. To convince the last of the doubters, Begin ordered Generals Eitan and Ivry to the meeting to present the IAF's secret plan of attack. To discuss the mission in detail, General Ivry brought with him Zeev Raz. Together, Ivry and Raz outlined for the ministers in precise detail the entire raid, from takeoff to return. The exacting specifications of the plan and its exhaustive attention to every detail clearly impressed the assembled ministers. Raz confidently answered every question and put to rest any doubts. At the end of the presentation, Begin called for a vote. The mission was unanimously approved by all ten ministers. Begin then set the day of the attack: May 10, 1981, a Sunday, seven weeks before the June 30 national elections.

"What are we calling the attack?" Begin asked Eitan at the end of the meeting.

The bushy-eyed chief of staff fixed the prime minister in his stare.

" 'The noise of battle is in the land, the noise of great destruction,' " Eitan recited, quoting Jeremiah from the Old Testament. " 'Before your eyes I will repay Babylon and all who live in Babylonia for all the wrong they have done in Zion, declares the Lord.' "

Eitan smiled thinly.

"We will call it Operation Babylon."

A week later, at Ramat David, Raz called the entire squadron into the briefing room. Nachumi, Yadlin, Yaffe, Katz, Shafir, Ramon, Falk, and Spector all took their seats.

French Prime Minister Jacques Chirac shares a laugh with Saddam Hussein during his first trip to Baghdad in 1974 to discuss selling Iraq a nuclear reactor. The two leaders would remain friends for nearly three decades.

Israeli Air Force Gen. David Ivry spent four years devising the precision attack on Iraq's Osirak reactor, the first time in history that a nuclear reactor was bombed.

Chief of Israeli Military Intelligence Gen. Yehoushua Saguy shocked Israeli Prime Minister Menachem Begin and the mission planners when he opposed any preemptive strike against the Osirak reactor.

IDF Chief of Staff Gen. Raful Eitan escorts Begin past the mission squadron's F-16s on a tour of Ramat David Air Force Base the day after the raid on Osirak.

Deputy Prime Minister Yigael Yadin takes the podium to celebrate the delivery of Israel's first four American-made F-16s on July 2, 1980. Yadin would later oppose using the planes to attack Osirak.

The sleek, technologically advanced F-16s flew just 100 feet off the ground through hot, unstable desert air for 600 miles to Baghdad in order to avoid enemy radar.

The mission's youngest pilot, Ilan Ramon, was later chosen to be Israel's first astronaut. He perished in the Columbia space shuttle accident in February 2003.

End of a dream: the ruins of Osirak and the surrounding al-Tuwaitha Nuclear Research Center marked the beginning of the end of Hussein's plan to make Iraq a nuclear power.

"I know you all have been wondering for a while what our target is going to be," Raz said evenly. "I can tell you now. On May 10 we will take off from Etzion Air Base and fly to al-Tuwaitha, a nuclear facility south of Baghdad in Iraq. There we will bomb the Osirak nuclear reactor."

The silence in the room was deafening, as Rani Falk would think of it later. Ramon, Nachumi, and Spector had long known the destination. Katz had figured it out, but the others were clearly stunned. Later, the men would contend that they had already figured out the target. But looking around the room, Falk saw by their stricken faces that no one in their wildest imaginings had guessed the target would be a nuclear reactor.

As Raz detailed the mission and discussed the formidable AAA gun emplacements and SAM batteries surrounding the complex, the many risks involved were plainly obvious to each pilot. The fact was in 1981, Israeli Air Force tactics and weapons systems still lagged far behind the efficiency of the new Soviet antiaircraft and SAM technologies. Syria's advanced, computerized, and radar-guided Soviet-made SAM-6s, which locked onto the exhaust heat of the fighters' thrusters, had shocked the IAF in the opening days of the Yom Kippur War. One Skyhawk squadron lost 17 of 30 planes to SAMs in two days over the Golan Heights. Yadlin's unit alone lost nine pilots—an unthinkable casualty rate before '73. In the past thirty years, dating from the beginning of the Vietnam War, 90 percent of all air force kills were due to AAA fire. Much of the reason was that the now widely used heatseeking SAMs were forcing pilots to take more and more dangerous avoidance maneuvers, ultimately causing them to accidentally wind up in the AAAs' deadly line of fire. The majority of the losses were between 1,500 and 4,500 feet—exactly the elevation of the pilots' tracking on final. The consensus was, you had no more than ten seconds to go low to low—that is, pop-up to release—or you were a dead man.

Worse, Mossad reported that in the wake of the Iran bombing,

Iraq, besides putting all units on full alert, had beefed up the number of AAA batteries and SAM emplacements surrounding Osirak. That intelligence and the images of gigantic towering tethered balloons ringing the fortresslike facility were hard to keep locked away in a pilot's mental box.

Indeed, Operations had secretly run out the risk assessment numbers and determined that, given the mission parameters, the probability was the mission team, due to either AAA fire or mechanical failure, could expect to lose at least two aircraft. Ivry had chosen not to pass that grisly statistic on, but word of the death math had leaked out nonetheless. The pilots did not need to be told what they were up against.

A few days after Raz's briefing, Ivry ordered Raz and Nachumi to Tel Aviv.

"We have heard from all the engineers and the experts," Ivry told them. "Now we'd like to know what you, the pilots, think of the mission."

"We will destroy the reactor," Nachumi said. Then, making it clear that there was really no room for discussion of personal safety, he added: "After that, what else matters?"

The ordnance crew technician checked his watch, then walked in front of the F-16, making sure Raz could see him from the cockpit, then, bending deeply at the knees, ducked under the plane's wing for a final look at both MK-84 gravity bombs. Each was secure in its release clips, fastened just in front of its tail fins, which served to stabilize the bomb and keep it from wobbling as it was lobbed forward in its downward arc to the target. The clips had to be sturdy enough to hold the two-thousand-pound bombs, which would be bounced up and down, along with the wings, during the rocky ride hundreds of miles through hot, unstable desert air. When the technician was sure the clips were secured properly, he pulled the metal safety pin from each bomb. He was amazed at how close to the ground the overloaded plane was. The intake manifold was barely twelve inches off the tarmac. Holding pins in his left hand as he ducked back out from under the port wing, careful to avoid the scalding-hot exhaust from the plane's tailpipe, the tech signaled "all clear" to Raz. He then jogged off the tarmac, where he joined the other ordnance techs, each of whom had performed the exact same maneuver on the F-16s. He did a last-minute check of the runway for any gravel or small obstruction that could be sucked into the manifold and destroy the engine. Theirs would be the final inspections. No sooner had the techs cleared the asphalt than the first two fighters were already barreling down the runway, picking up speed. From now on, the mission was beyond the immediate help of Etzion.

CHAPTER 5

WHEELS-UP

A man's character is his fate.

— HERACLITUS

Heading back after the security meeting at the prime minister's office that March, Ivry's driver took the main road out of Jerusalem, down the winding brown mountain pass dotted with green Jerusalem pines that thrived in the cooler air of the elevation—a phenomenon that surprised first-time visitors to Israel expecting to see nothing but desert and wadis. Deep in thought, Ivry stared out the car window as they passed the charred, red-rusted ghosts of vintage lorries and jeeps junked by the side of the road, silent sentries that lay where they had been shelled nearly three decades before in 1948—monuments to the first Israeli soldiers who had tried to fight their way up the mountainside under murderous artillery fire from Jordan's crack Legionnaire Brigade in their doomed attempt to capture the ancient Hebrew capital. That attack had failed in the end.

Ivry's could not.

But within weeks the mission was threatened yet again. Israel's neighbor on its northern border, Lebanon, had been unraveling into an anarchic feudal battlefield of warring strongmen and radical ethnic Sunni, Shia, Maronite, Greek Orthodox, and Christian Druze factions all protecting their own business interests and territories—and all kept in check by neighboring Syria, which considered the country its de facto client state. Adding to the chaos, PLO leader Yasser Arafat and twenty thousand PLO fighters had moved into Beirut after being chased out of Jordan. Begin's government, pressured by right-wing hard-liners, began funneling support to Lebanon's aristocratic leader Bashir Gemayel and the militant Christian Phalangists, hoping moderate Maronites would unite the country under a more benevolent and Israeli-friendly stewardship. The Sunnis, backed by Syria, rose up immediately, sparking civil war. The Phalangists appealed to Israel for support.

In February, Syria deployed SAM batteries into the Bekaa Valley to support thirty thousand troops, well within range of the Israeli border. Israel demanded that Syria pull out the SAMs or the IAF would take them out itself. Skirmishes followed. U.S. special ambassador Philip Habib, a Lebanese by birth, negotiated a fragile peace that lasted until April. It was then that Syria refused to pull back its batteries from the Bekáa, and Begin laid down an ultimatum to either evacuate the SAMs or Israel would bomb them. Egyptian president Anwar Sadat, scheduled to meet with Begin at Sharm al-Sheikh as part of the Camp David agreement, was anxious to use the May meeting to diffuse the Syrian-Israeli conflict. Unfortunately, the day of the historic meeting was May 10, the day of the Osirak attack.

Israel, with a bank account of goodwill in the U.S., well-organized and powerful lobbying organizations, and a strong moral argument, figured it might get away with an attack on Osirak or the Bekáa. But no one in the government believed it could get away with attacking *both*. The prime minister called another emergency cabinet meeting. Once again Ivry saw his mission on the cutting block. Begin was forced to choose between two evils—Syrian missiles or Iraqi nukes. In the end Begin decided there was no choice, really. They would deal with Lebanon later.

"We will destroy Osirak," he declared. "We must delay this Satanic plan for years to come."

The weekend of May 10, 1981, was unusually busy inside the walled compound of Etzion Air Force Base, some twenty miles inland from the Israeli resort town of Eilat. Most Israeli soldiers were routinely given the weekend off to observe the Sabbath. But this weekend all leaves and passes at the base had been canceled. The base's telephone lines, with the exception of key operational communications, were cut off by order of the army's Security Field Service.

Altogether, fifty aircraft, CH-53 helicopters, and hundreds of troops were mustering for the attack, though only the mission pilots and high command knew what the target was. The CH-53s carrying combat search-and-rescue crews would take off an hour before the F-16s and then hover over the eastern borders of Israel for the remainder of the mission. Eight F-15s would fly support: two two-man F-15s to follow behind the F-16s and circle high above Saudi Arabia while serving as communications relay stations, and six F-15 fighters from Squadron 133—two to provide radar jamming above the target area, and four flying at a high altitude and distance to provide air-to-air combat support if needed.

Early Sunday afternoon, in the huge hangar, hundreds of technicians were checking and rechecking the twelve F-16s, loading ordnance, and affixing air-to-air Sidewinder missiles beneath the wings. At the door of the base squadron room, dozens of F-16 and F-15 pilots, crew members, and commanding officers, having completed final briefing, were climbing the stairs from the briefing room to ground level. The men's faces were taut, expressionless. There was no small talk. Rani Falk was as keyed up as the others.

But his excitement was tempered with disappointment. After all, he had trained for this day alongside the other men from the beginning. Until recently he thought he would be going with them. Now he was a backup pilot—on hand only for the most dire emergency. He was not angry or resentful. But he felt let down. He had to keep himself ready, though. You never knew. He would fire up his F-16 and taxi it out of the hangar just like the others.

A Mossad officer, his right hand clutching a black briefcase, approached the group of pilots outside. Suddenly the snap of his briefcase popped and the bottom fell open, spilling out thousands of dollars in Iraqi dinars. The bills blew down the runway past surprised mechanics and passersby. The dinars were for the pilots in the event someone was downed behind enemy lines. The money could be used as a bribe for safety. The officer hurriedly gathered up the bills, too panicked to be shamefaced. Amos Yadlin smiled grimly. So much for any doubts about the destination of their top-secret mission.

The time just before takeoff was the worst. Once in the air the men would be busy with the job at hand. But waiting around . . . the thoughts began to creep in. The mental box opening ever so much. The doubts, though, were a fear of making a mistake, of somehow letting the team down. Few thought about personal safety.

Second team leader Nachumi had gone ahead of the others to start his plane's engine and begin checkoff procedures with his maintenance chief. He was running over the switches, checking navigation, weapons, electrical. He looked up to see his chief standing in front of the plane gesturing to him. He was slashing one finger across his throat. Nachumi could not believe his eyes. That was a kill gesture.

"What?" he yelled into his helmet microphone over the deafening whine of the Pratt & Whitney.

He was wound up like a spring.

"It's a 'No go,' " came the crew chief's response.

"Shit!" Nachumi exclaimed.

The pilots were given the news: The prime minister himself had cancelled the mission. It took a moment for what had been said to sink in. They were stricken. Yadlin was pumped with adrenaline, his mind and body intensely focused, coiled—as though he were in another dimension. Now suddenly the air was let out. He felt as though he had been stabbed. Spector, who'd had missions scrubbed before, immediately worried that this was it—the raid might be called off for good.

The hundreds of troops stood down, the F-15 pilots were re-called to base, the helicopters sent home. Hundreds and hundreds of man-hours, hundreds of thousands of dollars, wasted. What the hell had happened? Ivry wondered.

Several hours earlier that Sunday morning, Begin had been in-terrupted at his home by a special courier carrying an urgent letter from Shimon Peres, the former defense minister and the Labor Party's candidate for prime minister. Peres had learned the date of the attack the previous evening, May 9.

Begin read the note.

May 10

PERSONAL—TOP SECRET

Mr. Prime Minister:

At the end of December 1980 you called me into your office in Jerusalem and told me about a certain extremely serious matter. You did not solicit my response and I my-self (despite my instinctive feeling) did not respond in the circumstances that then existed.

I feel this morning that it is my supreme civic duty to

advise you, after serious consideration and in weighing the national interest, to desist from this thing. I speak as a man of experience. The deadlines reported by us (and I well understand our people's anxiety) are not the realistic deadlines. Materials can be changed for materials. And what is intended to prevent can become a catalyst.

On the other hand Israel would be like a tree in the desert—and we also have that to be concerned about.

I add my voice—and it is not mine alone—and certainly not at the present time in the present circumstances.

Respectfully,

Shimon Peres

The letter, stiff and awkwardly worded, was purposefully oblique in case it fell into the wrong hands—especially the Israeli press. "Material" that could be changed referred to the so-called caramelized uranium. "What is intended to prevent can become a catalyst" reiterated Peres's fears that bombing Osirak would only intensify Arab efforts to achieve a nuclear bomb. "Present time" and "present circumstances" referred to Peres's conviction that the French elections, being held that very week, would sweep his close friend, socialist François Mitterrand, into the presidency. Mitterrand, far ahead in the polls, had openly opposed Chirac and Giscard d'Estaing's decision to supply Iraq with uranium.

Peres pleaded for Begin to delay the raid. In truth, his letter had already accomplished that purpose. Top-secret plans had been leaked, the attack compromised. Deputy Prime Minister Yadin, convinced Baghdad would sniff out their plans, had earlier warned the cabinet: "They're ready for us." That prediction could be all too true now. Begin had been convinced all along that Labor would find some way to sabotage any raid on Osirak. Peres's claim that his objection was only to the "timing" was a red herring.

"Mark my words," Begin had told Sharon after the March cabinet meeting. "They would never accept such a decision. All the responsibility of doing this will be ours."

Begin was furious. Who had leaked? And how many others knew? He called Eitan, Sharon, and Shamir. All agreed: the mission had to be scrubbed. Begin called Ivry at IAF command at Etzion and ordered the pilots to stand down.

He would bitterly resent Peres's May Revolt for the rest of his life. He set about immediately to discover the source of the leak, the "betrayer." Though it could not be proved, Begin and his supporters were convinced that former defense minister Ezer Weizman had tipped Peres and Labor leader Mordechai Gur the day before. Weizman, in on the planning from the earliest days, had bitterly opposed the attack. He had many friends within the military and in Begin's government. Learning details of the attack would have been easy for the politician.

Begin vowed not to make the same mistake twice. From that day on, he announced, any decision on an attack on Osirak would be made by just three men: himself, Foreign Minister Yitzhak Shamir, and Agriculture Minister Ari Sharon.

The following week the French duly elected François Mitterrand their president, and he indeed responded to Israel's objection to France's nuclear treaty with Iraq. France would no longer engage in the sale of nuclear technology to Iraq, Mitterrand declared. But, regrettably, the country was bound to honor its present agreements with Saddam Hussein. Iraq would receive full delivery of all seventy-two kilos of enriched weapons-grade uranium.

According to Mossad, it already had.

A week later Begin met secretly with Sharon and Shamir. Eitan and Ivry were informed that the attack was set for Sunday, June 7, 1981. Ivry felt a great weight taken off his chest. For months, week after week, he had been edgy, ping-ponging mentally: Was the attack on or off? Would this be the Sunday?

The friction between the two team leaders, Raz and Nachumi,

rubbed even rawer under the stress of waiting. Indeed, Raz had begun picking up strange signals from Nachumi. Two days after the
mission had been canceled, back at Ramat David, he had been
called to Spector's headquarters. Raz walked across the base and
quickly mounted the few steps into the long barrackslike structure
housing the command and administrative offices. He passed various open doors and entered the outer vestibule of Spector's headquarters. A frequent visitor, the squadron leader did not bother to
wait for a formal announcement of his arrival. Instead, nodding to
the secretarial staff and assistants manning the desks, most of
whom were young college-bound kids doing their mandated military service, Raz strode directly into Spector's inner office and was
surprised to discover he already had a visitor. Nachumi stood beside Spector's desk, leaning in toward the commander. The two of
them had been talking softly, careful that their voices did not carry.
As Raz entered the room, Spector and Nachumi looked up surprised—and, Raz realized—embarrassed. Neither man had expected to see Raz, that was clear. Raz knew immediately he had
interrupted a very delicate conversation. He knew also, with a certainty he might not have been able to readily explain, that the two
pilots had been talking about him. An awkward silence hung in the
room for an uncomfortably long time.

Raz said nothing.

Colonel Spector recovered first.

"Well," he said, continuing in a conversational tone as though
Raz had been in on their chat all along, "Raz has been squadron
leader the whole mission, so he will stay leader."

Nachumi nodded, glanced awkwardly at Raz, and left the room.

Raz stood rooted where he was. What the hell was Spector talking about? Of course he was squadron leader. When was the notion that he would continue as squadron leader ever in doubt?
Spector spoke as if it had been somehow debatable—and out in
the open! Raz opted against pressing the matter and let the inci-

dent pass as some kind of mixup. But privately he was enraged. He could not believe the treachery he had just witnessed. Nachumi, at the last minute and behind his back, had seemingly been lobbying Spector, his mentor, to promote him to mission leader over Raz. Forget the question of insubordination, or that it was General Ivry's call to decide who was in command, on strictly professional grounds, who would risk the cohesiveness and camaraderie of a combat team facing a formidable assignment on the eve of the mission—for personal advancement? Indeed, Raz perceived, Nachumi considered Raz and himself co-leaders—it was simply a matter of timing that Raz was first team leader. As he had said, missions in IAF were assigned to squadrons, not a single man. Raz could not believe the gall.

But Raz put it away in the box. He would not tell anyone on the team what he had witnessed—it would risk the same damage and resentment throughout the entire squadron that he alone was feeling now.

On Wednesday, June 3, four days before the attack, Yoram Eitan took off in an Israeli-made Kfir fighter for a training exercise in air-to-air combat in the skies above Etzion. Doobi Yaffe had been Yoram's instructor when he was still a "nugget." Yoram was not a natural pilot, but he was determined, full of energy, and fearless. And a bit headstrong. During the mock dogfight, Yoram made a radical maneuver to evade his "attacker," pulling the plane's nose up to escape the enemy on his tail. The engines did not "flame out," but the Kfir began stalling, losing airspeed quickly. Suddenly the plane rolled over and began a flat spin, gyrating violently, turning round and round, corkscrewing toward the desert floor. Yoram worked desperately to regain control of his aircraft, fighting the G forces that pinned him to his seat and made the simplest of movements extremely difficult.

"Neutralize the stick!" the OT instructor radioed Yoram.

"I'm, ah . . . ah . . . trying," Yoram gasped, fighting the spin and the dizziness.

With each revolution his plane lost a thousand feet.

"Come on, Yoram. Get it stopped!" the instructor yelled through his oxygen mask.

"I can't . . . ah . . . I . . . it won't stop . . . ah . . . spinning . . ."

"Get out! Yoram!" the instructor pleaded desperately. "You're getting low. Eject!"

"Wait, I think . . . I think . . . I . . . ."

"Eject. Eject now!"

Yoram's Kfir continued to corkscrew horribly toward earth. Inside the aircraft he was disoriented, fighting unconsciousness, struggling hopelessly with instruments that would not respond.

Then, no more radio contact. Just a terrible silence. And the telltale black plume of smoke rising from the pale desert floor below.

As Eitan sat in the meeting at IAF headquarters, a pale, grim-faced aide entered and crossed the room to tell the general the news, his voice little more than a whisper: his son, Yoram, had just been killed in a training accident over Etzion. Eitan swallowed the news, saying little. He left the briefing room and drove straight home to his wife, where the two parents began sitting shivah for their lost pilot.

Word quickly spread throughout the military. Every pilot at Ramat David was shaken. Yet another reminder that death could come for anyone at any time. But mostly they ached for Eitan, who had always been more like a beloved uncle than their commander and chief. No one said a word, but everyone had the same thought: How bizarre was it that the death of their commander's son should take place at the very base they were to take off from on the most desperate mission of their lives? What did it mean?

Ivry grieved for his friend. But he had another problem as well. He had been invited by the United States Navy to attend a gala celebration in Naples, Italy, the weekend of the 7th to mark the change of command of the Mediterranean's 6th Fleet. The U.S. Navy would fly Ivry into Naples Thursday night, fete him at Friday's reception and black-tie dinner, then fly him home Saturday morning. The invitation could not have come at a worse time. The outgoing admiral was a longtime friend of Ivry's, so to beg off would be personally offensive. But, more seriously, considering what was going to happen Sunday, a no-show at the function, in hindsight, would look almost like a betrayal to the Americans. On the other hand, to show up, knowing all along that his own air force was about to violate an arms treaty between the two countries, and to eat, drink, and chitchat at his allies' expense on the eve of his "perfidy," as he was sure some would see it, might look like he was rubbing it in. And, of course, while trying to be diplomatic and charming, his mind would be elsewhere, worrying about details of the mission. The flight to Italy was torture as all these thoughts and more ran through Ivry's head.

Eitan was supposed to get a final go-ahead from the prime minister on Friday, June 5. Ivry ordered his aide to call him in Naples as soon as he received word: if the raid was still a go, he was to use the code word *Opera*. Sure enough, late in the afternoon at the gala, as Ivry made small talk, chatting up the navy brass and nibbling without appetite the appetizers served in a grand ballroom overlooking the most beautiful harbor in the world, the IAF general was called away to the telephone.

"Yes?" Ivry said into the phone, his muscles tensing unconsciously.

"Your tickets for the Opera have been confirmed for Sunday," he heard a familiar voice say from Tel Aviv.

"Thank you," the general replied, and hung up.

Ivry didn't know whether to feel relieved or anxious. He still didn't know how to feel Saturday morning, when he arrived back home to discover that his wife was having guests in for dinner that night.

The question of whether or not to tell their wives about the mission weighed heavily on all the men—with the exception of bachelor Ramon. He had been dating a pretty aide in military intelligence named Ophir, and she knew all about the raid. Most of the other pilots decided not to tell. First of all, they had all signed a paper swearing not to reveal details of the mission to anyone. Second, and most important, the pilots did not want to put their wives through the torture of waiting and worrying. What would be, would be. Worrying them to death was not going to change anything. It was especially tough on Raz, whose wife had just given birth to his son.

Doobi Yaffe's mother, Mitka, not only knew about the mission, but had known about it long before her son had heard of Osirak. She knew everyone who was anyone because her husband and Doobi's father had been a famed IAF commander, but also because she had served as personal stenographer to every prime minister of Israel dating back to David Ben-Gurion. She was now Prime Minister Begin's right hand and had been at every cabinet meeting, taking notes on every detail of every briefing. Mitka would be with Prime Minister Begin on Sunday, awaiting word of the attack. Two days before the mission, on her way to see Raful Eitan in Tel Aviv, Mitka stopped by to visit her son and his wife, Michal. Michal was the daughter of Ezer Weizman, who was not only the former defense minister, but had had a famous falling-out with Doobi's own father, Avraham Yaffe. The two celebrated military heroes had been best friends—that is, until 1968 when

Weizman, then commander of the IAF, refused to recommend Avraham to succeed him, launching a feud between the two well-known Israeli figures that lasted a decade, during which neither would speak to the other. As a result, Doobi and Michal's unlikely courtship and marriage was something akin to the one between the Capulets and Montagues.

Yaffe had told his wife about the mission. Mitka Yaffe could see it at once in Michal's pale expression. The two women never exchanged a word about the impending raid or hinted that they were aware of it. But as she was leaving, Mitka hesitated at the doorway and locked eyes with her daughter-in-law, holding her there.

"When he is back safely," she said deliberately, "I will call you and tell you to have a glass of cognac. And you will know . . ."

Hagai Katz had made up his mind not to tell his wife, Ora. But earlier that first week in June, just nights before he was to fly to Etzion, Ora informed Hagai that they were invited to a family gathering that weekend with her parents.

"I can't," Katz said. "I have an important mission this weekend."

"You have to come," Ora flashed.

"I cannot," Katz repeated. "It's very important."

"Oh, sure," Ora replied sarcastically, convinced Hagai was trying to wriggle out of the dinner with her in-laws. "What are you going to do, bomb the Iraqi nuclear reactor or something?"

Katz nearly fell backward. He stared at his wife, speechless. But thankfully, she was much too annoyed to notice.

Amos Yadlin trusted his wife Karen more than he trusted himself. The two shared everything, even more so after she gave birth to their first daughter in February. So one night before the mission, as the team tied up loose ends at Ramat David, preparing to fly the F-16s south to Etzion on the fifth, Yadlin sat Karen down and told her about the mission he was to undertake on Sunday. He did not need to tell her how dangerous it was.

When he finished, Karen stared at him, her eyes shiny, filling. But she was not going to cry. She took his hand in hers and squeezed it.

"Try to survive," she said quietly. And that was that.

The Friday before the mission, the pilots flew their F-16s down the Sinai to Etzion one at a time, staggering their flights throughout the day in order not to attract attention. Raz took off from Ramat David in his fighter, No. 107, in the early afternoon for the one-hour flight to Etzion. The centerline tank and the two wing tanks were empty, but he carried two Sidewinders. As he flew above the country, he watched the land below change from verdant towns and kibbutzes to the brilliant colors of the Negev. But what should have been an easy flight was becoming surprisingly laborious. Raz felt that his INS, the inertial navigation system that automatically computed and adjusted the preflight navigation headings, factoring in air-speed and mileage as well as wind and weather, was not accurate. "Washy" was how he thought of it. When he landed, he complained to his wingman, Amos Yadlin.

"It's not good," Raz said. "As the leader, I need a better airplane."

"Take mine then," Yadlin said.

In truth, Yadlin thought Raz's complaint was somewhat odd. Training and flying in the same plane for months, a pilot became attached to his aircraft as though it somehow had its own personality, its own "soul." Indeed, like high-performance automobiles, each plane handled a little differently, had its own quirks and mechanical signatures a pilot grew used to. To simply switch to a new aircraft voluntarily, especially at such a late date, was almost unheard-of. But Yadlin knew the squadron leader was under a great deal of pressure. Yadlin liked his plane, No. 129, but he wasn't overly superstitious about it, as some pilots were. He was too practical for that. If it would make Raz feel better, then he was willing to switch. It was decided: Raz would fly 129, Amos 107.

The rest of Friday afternoon the pilots were left mostly to themselves to read, relax in their barracks, or wander the base. Meanwhile, the tech crews and mechanics checked out the twelve F-16s sequestered in their hangar. The familiar blue six-point Star of David and white circle on the wings were painted out with the sand, brown, and green iguana pattern of desert camouflage. The MK-84s, external fuel tanks, and everything else was checked and rechecked.

That night, after all the pilots had landed and been billeted in the officers' barracks, the men watched a 16mm Israeli war movie in the squadron room and then turned in to bed early in the large dormitory-style room.

Saturday morning, the pilots were up by 0600 simply out of habit. It was to be a day of so-called leisurely activity—a torture for soldiers waiting to attack. The pilots were used to short-time combat. You were up, you spotted an enemy MiG, you got cleared to engage, and then you engaged and killed—all in less than ten minutes. Or in combat, you flew sortie after sortie, landing, refueling, rearming, and taking off again. In both cases Israeli pilots had no time to think, to ponder the unknowns and the what-ifs. This mission was an entirely new experience for the IAF, and it brought new problems. The squadron had just discovered the latest: time. It was their own personal version of *Ha-Hamtana*, literally "the waiting," what the Israelis called the tense, maddening, interminable period of time in 1967 after Egypt reoccupied the Sinai and the nation waited for the inevitable war to follow.

The men picked at their breakfast without enthusiasm. That afternoon they organized a basketball game in the base gym, Raz's squadron 117 team against Nachumi's 110. The friendly game, however, quickly turned competitive. They were all young, aggressive fighter pilots and no one liked to lose. The play grew rougher and more serious. Relik Shafir was a great shooter and drew a crowd of defense from Raz's 117. Soon elbows were being thrown, then body blocks and head butts. Shafir was knocked to the ground and a scuffle broke out under the basket. Some of the unspoken rivalry between the two leaders, perhaps, and maybe lingering resentment over Spector's gambit to join the squadron, had found its way into the game. Driving to the hoop, Yaffe was knocked hard to the floor under the basket, almost cracking his head open against the post. He climbed back to his feet, surprised. The fall snapped the men back to reality. The two teams decided to call it a game and headed back together to the dorm to shower and cool off.

Saturday night, as they lay on their bunks after lights-out, sleep came to no one. Ivry's wife, Ofera, a hundred miles away to the north, wasn't the only one tossing and turning. Staring up at the ceiling in the dark, each alone with his thoughts and trying not to think about tomorrow, the pilots began to joke and kid one another to relieve the tension.

Yaffe piped up from the dark.

"Ilan, you know, chances are, one of us will be staying in Baghdad," he said in mock gravity. "We talked it over, and we decided it was you."

"Why?" Ramon cried.

"Because," Yaffe replied.

"Because what?"

"Well, you're the youngest, you're the only one not married, you're the only captain, and you're number eight," Yaffe said matter-of-factly.

Ramon was speechless. This was his first combat mission. Yaffe's hazing was almost cruel. Complete silence filled the room. And then the pilots broke out laughing. It was the first real release of the pent-up tension of these last days.

"Hell with you," Ramon groused. "I bet we *all* come back."

"You're on," Yaffe snapped, not thinking about what he was saying.

The men erupted once again. After a minute or two, Ramon spoke for the last time that night.

"That's one bet I hope I collect."

Sunlight from the first blush of dawn in the east crept across the Negev and broke through the barracks windows, eating away at the last shadows of nighttime lingering in the corners of the room. Just five o'clock, it was already warm in the desert as the first sounds of life stirred outside. In the distance, a truck engine turned, the gears grinding. Sunday, June 7, 1981, had dawned.

Raz lay on his back, hands behind his head, staring at the ceiling. He had been awake for a while. None of the men had slept well. There was little talk as they stirred and fell into their morning routines—showering, shaving, pulling on their air force fatigues, and lacing up their boots. Again, the pilots had no appetite for the breakfast of rolls, fruit, and coffee set up outside the dormitory. A final briefing was scheduled for noon. Until then, they were on their own.

The adrenaline level was palpable in the room. The mess steward could feel it when he came to refill the coffee urn. The pilots spent until 1100 picking at the food and drinking tea or coffee, each man individually looking over the latest intelligence reports again, the maps, flight plans, and lists of code words. Everyone had long ago committed them to memory, but there was no room for error. And besides, it was something to do. They organized their notes and flight plans and navigation calculations, then clipped them to their kneeboards, the compact, hardbacked clipboards that fastened to their thighs so they could quickly flip to the pages they needed during flight. Strictly speaking, the kneeboards carried in the planes were frowned upon. Flight commanders warned that if forced to eject, the boards could become entangled in the seat harness or the ejection seat itself and rip their leg off. Some of the pilots had arranged special places in the cockpit to hang their boards so they could access them easily. This was also against regulations, since the boards, during an ejection, could become dangerous missiles in the updraft. But most pilots routinely ignored the regulation.

The pilots also spent time packing up their personal effects. These would be transported back to Ramat David later in the day. Hopefully, their wives would not be forced to retrieve them later.

They were served an early and large lunch. The men forced themselves to eat, even though no one was hungry. They knew they would need all the nutrients and energy they could get in the next seven or eight hours. At 1200 the men filed past the beefed-up security detachment outside the squadron room, the guards toting M-16s as they stood alert to any movement. The pilots carried their kneeboards and pencils to take notes. Inside, they took seats in the rows of armchair desks. Only those personnel directly involved in the raid were to be briefed. That included the F-16 pilots, the six F-15 support pilots, the two F-15 pilots and their radiomen who would form the communications link, and the briefers from Operations. The mission commanders were also there: General Saguy, IAF head General Ivry, and Avi Sella.

And standing next to Ivry was the chief of staff, Raful Eitan. The men were stunned. He was still sitting shivah. They had not expected him to be there. Eitan met their looks. He was unshaven, his eyes red-rimmed and ringed with dark circles, but his uniform was immaculate, his shirt laundered and starched. The briefing began and everyone took their seats. The weather report was first: clear with some cumulus clouds over the mountains to the east. The desert air would be hot, making for a bumpy ride in at only fifty feet off the ground.

"You new guys, take your airsick bags," the briefer quipped a little stiffly.

The men tried to force a laugh. The tension in the room seemed to suck the energy out of everything. The briefer also repeated the intelligence that because it was Sunday, the foreign workers would all be at home.

"What if they're not?" Shafir asked.

Ivry stirred in his chair.

"We didn't ask them to be there," he snapped. His eyes burned angrily. "We warned them and their leaders many times to go home. If they don't want to, they are on the side of the terrorists, and whatever happens to them is their choice."

Saguy, as head of army intelligence, went over the enlarged ground photos of Osirak and the al-Tuwaitha complex taken by Mossad, pointing out their target, the dome of the reactor. He reviewed the intelligence reports on the latest placements of AAA and SAM batteries as well as the SAM-3s and SAM-6s that fanned out all the way to Baghdad. Intelligence also reported a new brigade of SAM-6s at the site. Each brigade included five batteries, each battery armed with twelve telephone-pole-size missiles carrying 145-pound warheads. That added up to sixty surface-to-air missiles to fire at the Israelis. This was new information, and unwelcome at that. There were photos, too, of the present positions of the mobile ZSU 23-4 radar-guided antiaircraft guns, which looked like missile tanks.

"Remember," Saguy said, "the SAM-6s are smokeless. You can't see them. And they fly twice the speed of sound. In our estimation, the SA-6 missiles pose the most dangerous threat."

Saguy then retraced the navigation route, which would dogleg south of Jordan from Etzion to avoid Jordanian radar and combat patrols, then cut across Saudi Arabia and through the western border of Iraq straight through to al-Tuwaitha. The Saudis had four American-supplied AWACS equipped with powerful long-range search radar extending as far out as 350 miles. IAF did not have the power to jam the AWACS. But intelligence reported that the Saudis would deploy only one AWAC during the mission parameters, and that plane's normal search pattern focused on the Persian Gulf to the south, not to the west.

"We think there is a hole, a strip of desert all the way to Baghdad, with no radar coverage by anyone," Saguy concluded. "But keep your head on a swivel, just in case."

Ophir, Ramon's girlfriend, covered the remaining intelligence details. When she finished, to the mild surprise of the assembled fliers and fighters, she blew him a kiss. He gave her a small wave of the hand, but it was not enough to hide the anxiety on his face.

Raz stood up from his desk and took the podium to review the entire mission—call signs, takeoff procedures, navigation routes, code words, radio frequencies, radio silence protocols, and emergency procedures in case of mechanical failure. The F-15s, billeted in the underground hangar, would taxi out for takeoff at 1455, or 2:55 P.M. Avi Sella would fly in one of the F-15s, operating a large, bulky SSB HF long-range radio. His job was to follow some twenty miles behind the attack force and function as the relay station between the mission leader and command.

A Boeing 707 communications command post had already taken off and would orbit above Israel. Sella in the F-15, following behind the attack group, would relay his radio messages to the Boeing communications post, which would in turn relay the messages to IAF control and command on the ground at Etzion. The F-16s would taxi out at 1500 hours for takeoff at 1600. The attack group should arrive at the target just before sunset, flying eastward out of a blinding sun. On their return, fighters would be scrambled over Israeli airspace to escort the F-16s home. GCI, ground-controlled intercept radars, would monitor any Jordanian activity.

They would fly in two teams led by Raz and Nachumi. Each pilot would have a call sign according to his place in the bombing order.

"I am Blue One," Raz said. "Yadlin is Blue Two. Doobi and Hagai, Blue Three and Four. Amir, you are Blue Five, Colonel Spector Blue Six. Relik and Ilan, Blue Seven and Eight. Any questions?"

The teams, Raz continued, would fly at 360 knots, or six miles a minute, then increase to 480 knots, or eight miles a minute, for the approach to target after passing the IP, or initial point, the final navigation point at which the bombing run would start. The IP, Raz reminded them, pointing to the map of Iraq that hung behind the podium, was the edge of an island in the middle of Bahr al Milh Lake, about four miles west of al-Tuwaitha. At that point the pilots should line up and fly with thirty seconds' separation between the planes. Iraqi radar would probably pick them up then, as they climbed out of the "snow," the clutter of low-elevation radar. At this time the F-15s, which had shadowed the attack group twenty miles behind, would close and flip on their search radar, hunting for MiGs as they climbed to umbrella the target area.

Raz looked toward the six F-15 pilots.

"Your job is to protect us from MiGs," Raz said. "We cannot be deterred by dogfights."

Zeev looked at his own team.

"Remember, if you have trouble and have to eject, climb no higher than one hundred feet," Raz reminded them.

The ejection charge would shoot the pilots to five hundred feet. Since Iraqi radar made sweeps only every twelve seconds, chances were a parachute drop from six hundred feet would show up as a blip only once before the pilot hit ground, so odds were good the blip would be read as a mistake.

"If you are hit by AAA and can fly, head west as far as you can. Drink your emergency water as soon as you get on the ground," Raz said. "It'll help you get over the shock of bailing out. Gather your parachute and bury it, then begin walking west. Wait until dark before activating your PRCs. We can't afford any strange events like a rescue beacon before the attack. The success of the mission depends entirely on surprise."

He stopped for a moment and shuffled his notes together, then looked back up.

"We have to destroy this target at all costs. There is no secondary target." Raz stopped again, and looked at Nachumi.

"This will be tough for you, Amir. You're used to being high up in dogfights rather than down below pounding sand like the rest of us," he smiled.

Nachumi and the pilots laughed.

"If I'm hit, Amir will be in command," Raz added, becoming serious again. "If Amir is hit, Amos is in command. After escape, once everyone has checked in, I will call headquarters and let them know how many of us made it."

He looked up from his notes and eyed the pilots sitting before him.

"So, if there are no other questions . . ."

Eitan stood up from his chair next to Ivry and walked to the front of the room. The men stirred in their seats a moment, then a deep silence filled the room as the general stood before them.

"This is an important mission, and a dangerous mission. I worry for your safety," he said, his voice clear and strong, but strangely soft. "If something happens, I want you to know that we'll do all we can to rescue you. Don't try to be some special kind of hero in the face of torture. Tell them what you have to. We want you back with sane minds. We understand what you'll be going through."

Yadlin realized the general was addressing them like a father. Eitan had just lost his son, and he could not stand the thought of losing another of his "children."

"Your government and the people of this country are appreciative of your efforts and sacrifice. Your willingness to risk your lives, so we might live, will never be forgotten by Israel. This is no ordinary mission. Never before has the Israeli Air Force flown an attack to such a distant point—and for such an urgent need. Our history as a nation and as a people is at stake."

Eitan looked at the faces of the pilots sitting before him, indeed, like schoolchildren at their desks.

"You've all read the Bible. You know the history of our people. You know how God brought Moses and the Jewish people out of Egypt. You know the battles Joshua fought to gain entry to the Promised Land. You know about the just rule of King David and the wisdom of Solomon. You know about the dispersion to Babylon. We've kept our identity as a people. And now, nearly two thousand years later, we are reunited as a nation.

"Our people have overcome the agony of the Holocaust. We've gone through a modern-day exodus. We've survived wars in 1948, 1956, 1967, the War of Attrition in 1970, and the Yom Kippur War. And now we are faced with the greatest threat in the long history of Israel—annihilation and destruction of our country with atomic bombs by a madman terrorist who cares nothing for human life. We must not allow him to achieve the ability to build the bomb that could destroy us.

"That's what this mission this afternoon is all about. Protecting our country. The future of Israel rests on your skill and ability to destroy that nuclear reactor. You must be successful—or we as a people are doomed. This is a pivotal point in the history of Israel. . . ."

Eitan raised his arm, his hand tightened into a fist. His voice now was strong, loud but without being raised.

"If we are to live by the sword, let us see that it's kept strong in the hand rather than at our throat!"

The room seemed to almost shake with the energy pulsing through the bodies of the men. One could almost hear a roar echo in the silence. Embarrassed by the hypnotic focus his words and passion had wrought among the assembled men, Eitan tried to defuse the tension, pulling out a bag of dates—a notorious staple of prison food in the Middle East.

"Here," he said, offering the official fruit of Iraq to the troops. "Have some of these. You'll have to get used to them where you're going."

The men, including the assembled generals, broke into laughter as each pushed forward to grab a date and share in this final "toast."

As the men began to gather their kneeboards and file toward the doorway, Ivry called out.

"God be with you."

The pilots suited up back in the barracks. Each man wore a lot of gear. First the flight suit, then the G-suit, the torso harness, survival gear, and finally, once in the cockpit, the helmet. For a normal flight the special suits were not a problem, but the pilots would be strapped into these uncomfortable combat clothes for nearly four hours. Sleeves that pinched under the arms or a collar that chafed at the neck could be painful and distracting after hours cooped up in a stuffy cockpit. Then each man grabbed two PRCs, emergency radios about the size of a Walkman that sent out a homing signal to the rescue choppers in case they were shot down. Normally a pilot took only one along. But there was nothing normal about this mission. No one wanted to take a chance of being stranded in the Iraqi desert with a dead PRC. The men clipped the radios to their torso harnesses and headed out to the four vans waiting to take them to the hangar.

Outside, the desert air was volcanic hot, the sun an angry white ball burning high above, bleaching the sky a desiccated blue. Raz stared down the runway to the hangar. Sheets of heat waves shimmered off the baking tarmac. The pilots carried their gear into the maintenance center, where crew chiefs and mechanics stole glances at them. Everyone knew that something big was going on, though no one knew what. The pilots signed for their planes and checked the maintenance reports. The F-15 pilots broke off, heading for the camouflaged hangar at the head of the runway that housed their aircraft. The conformal fuel tanks had been bolted onto the fuselages, and each plane was armed with four heat-seeking Sidewinders and four radar-controlled Sparrows as well as some five hundred rounds of 20mm cannon fire.

Raz's group passed the open-ended green hangar where the four backup F-16s were parked and headed to the eight armed and waiting planes each man had been flying for the last six months—with the exception of Raz and Yadlin, who had swapped. The aircraft had been moved out of the underground hangar and into the hot sun—exposed to the prying eyes of American and Soviet spy satellites orbiting high above the earth's atmosphere—to avoid any maneuvering and sharp turns around corners. The planes were thousands of pounds overweight and Operations feared the extra pressure could collapse the landing gears.

The pilots climbed out of the vans and walked around their planes checking for hydraulic or fuel leaks, the bomb attachments, tire pressure, and any nicks or dents that might have gone unnoticed. At the tips of each wing were affixed the lethal heatseeking Sidewinder-9L air-to-air missiles, the most up-to-date models that tracked not only the heat from the exhaust systems but the heat caused by the friction of the aircraft flying through air. The pilots couldn't help but notice that their aircraft looked, somehow, well, *fat*. Indeed, sitting on the hot taxiways, the F-16s looked like over-loaded beasts of burden sagging under their lots. Katz worried: Could they really get them off the ground?

Finally, one by one the pilots climbed the metal ladders into the cockpits. Raz's plane was in the lead. His initial business was to settle in, try to get comfortable. He attached the parachute risers from his ejection seat to his body harness. Next, he plugged his G-suit ring into the cockpit air pump to blow up the protective bladders, which functioned somewhat like automobile airbags. He then snapped on the survival kit, complete with 9mm handgun, first aid, extra water, food packets, bandages, pain pills, even shark repellent—a holdover from standard World War II British protocol. He buckled his seat belt and switched the IFF (identification friend-or-foe) to standby mode, so the radio would not emit any electronic signal and give him away. Then he looked up. The crew chief straddling the ladder handed him his helmet.

"I don't know where you're going, sir. But good luck," he said, slapping Raz on top of the helmet and closing the canopy.

Lined up behind Raz and Yadlin, Doobi Yaffe cranked his air-conditioning all the way up. Sitting in the sun under the glass canopy was like being in a greenhouse. The air-conditioning struggled to defeat the rising temperature inside. Behind Yaffe, Katz pulled on his fireproof gloves. He had cut the tip off the index finger of the left glove so he could see his fingernail. That way he could double-check his oxygen system. If he began suffering hypoxia, or lack of oxygen to the brain, a subtle condition that creeps up on a pilot gradually, making him drowsy and interfering with his reasoning abilities before causing unconsciousness, his fingernail would turn purple.

This was nervous time. Once airborne, instinct and training would take over, forcing out any doubts. Each man had his own way of dealing with it. Spector spent the time looking over his maps, going over the mission in his mind. He thought of it as similar to being an actor on Broadway waiting behind the curtain for his cue. When the curtain went up, there he was: ready to take the stage. As he waited, Shafir considered for a moment his only real fear: not letting the team down, not making the big mistake. As for being killed? Well, that was out of his control.

Like the other pilots, Amir Nachumi was busy running through his checkoff procedures with his crew chief. When he hit the switch to check the electrical system, there was no response. He couldn't believe it. He flipped the switch again. And again, nothing. He began flipping more switches, rechecking gauges. The plane's electronics system had failed, including the INS navigation and threat warning. Nachumi called his maintenance chief.

"I'm going to kill you," he snapped, completely frustrated.

"What's wrong?"

"I got a no-go on the electronics," Nachumi said. He was angry.

"Try it again," the chief said.

"No go."

He had no choice: he would have to change planes. Unlike Raz, who had so cavalierly switched planes with Amos Yadlin, Nachumi was far more traditional in his attachment to his aircraft. After spending hours and hours in the fighter, depending on it literally with his life, he had begun to think of it as alive. Now the plane he had trained in for months and learned to depend on had failed. What did that mean? Nachumi worried. Was this a good sign or a bad sign?

Nachumi, like so many fighter pilots, was superstitious. Indeed, Spector and Katz had at first balked at taking part in a preflight group photo, thinking it might be bad luck to tempt fate. Nachumi's mind raced. Ten minutes before takeoff, he would have to get used to an entirely unfamiliar airplane. The bomb-sighting would be slightly different, the plane would handle differently in the air. Shit, he thought. He unbuckled and popped the cockpit, a furnace blast of Negev air pummeling him. He shimmied down the ladder and ran to the Quonset hangar to requisition a backup plane and begin preflight checkoff all over again.

Roaring down the runway, the F-15s began taking off in pairs, the deafening roar of their engines thundering across the tarmac, shaking the ground beneath the maintenance crews. Raz watched

Nachumi taxi the new plane onto the runway. He was not happy about the mechanical glitch, but what were you going to do? Things happen. Hopefully, this would be the worst. Once Nachumi was back in rotation, Raz taxied the F-16s to the beginning of the runway, then halted the squadron. The aircraft formed two staggered, diagonal lines. Four tanker trucks pulled up to the planes from the shoulder of the runways, careful to avoid the exhaust of the thrusters. Using hand signals and wearing protective earmuffs, the fuel crews climbed up the wings, dragging the hoses with them. As the planes idled, the crews began the "hot refueling," topping off the fighters, which had already burned up some four hundred pounds of jet fuel during checkoff and taxiing. A precious eight minutes of flying time.

The crewman outside Yaffe's plane was having a hard time. For some reason the fuel was not transferring from the tanker hose into the tank of the F-16. The crew chief signaled some mechanical glitch. Yaffe grew anxious. The extra fuel could mean the difference between landing safely back at Etzion or flaming out somewhere over the Saudi desert. I didn't train an entire year to turn back now, he thought. The hell with it. He waved off the ground crew. Raz looked behind him through the glass cockpit. Yaffe gave him a thumbs-up along with the rest of the pilots.

Raz checked his watch: 1557. He pointed his forefinger down the runway. The ordnance crews pulled the safety pins on the MK-84s, the ground crews, hunched over and holding on to their caps, circled beneath the planes for one final inspection, then Raz and Yadlin began taxiing slowly to takeoff, being careful not to put needless stress on their landing gears. Any sharp dip or angle and the landing struts could simply crumple. The rest of the group followed the two lead planes.

Raz pushed the throttle forward all the way, the asphalt beneath him becoming a blur as the fighter picked up speed down the runway. He shoved the stick into afterburner, heard the engine whine and roar, a plume of exhaust shooting behind him. He passed the 1,000-meter marker. His airspeed was 90 knots. The 2,000-meter mark flashed by. He was at 124 knots. On a routine mission he would be lifting off now. But the landing gear hugged the ground.

I'm too slow, he thought.

He needed to make 180 knots in order to get airborne.

Then 3,000 meters; 4,000 meters. He was still at only 145 knots. Raz's stomach tightened. He could see the 5,000-meter marker racing toward him. He was running out of runway. He cursed the extra weight. It would have been better to skip the extra fuel and risk flameout and ejecting over the desert than wind up a pile of charred, twisted metal at the end of the runway. He eased the nose back a bit. The plane seemed to slow for a nanosecond, then it thundered forward, finally lifting off at 5,200 meters. His airspeed indicator read 180 knots as the fighter climbed into the blue sky, already leaving the ground shrinking behind him.

Raz untensed. He looked to his side. Yadlin was there, just off his wing. He began a long, slow, banking turn southeast, leveling off for the "running rendezvous," the rest of the planes already dropping into group formation behind him. They were in a spread formation abreast, boxed in pairs: Raz and Yadlin, Yaffe and Katz, Nachumi and Spector, Shafir and Ramon. The eight fighters screamed east toward Aqaba, tickling the tops of electrical poles, one hundred feet above the ground, heading toward Baghdad and destiny.

Major Rani Falk watched the last of the eight attack planes go wheels-up through the glass cockpit of his F-16, idling at the head of the runway, ready for takeoff had he been called. He had watched the six F-15s flying support and two two-seater F-15s flying Com disappear into the southern skies an hour earlier. In all, sixteen aircraft. It was a bittersweet moment for Falk as he watched the fighters bank south, following in the direction of the F-15s. He felt happy that the mission was finally under way after years of training and preparation, and relieved that his squadron mates had successfully lifted off while flying dangerously over-weight. But at the same time, Falk found himself fighting an unde-niable disappointment, a longing to be with the fellow pilots he had trained with day in, day out, month after month, for two years.

It was only the luck of the draw—and perhaps some late-in-the-game political maneuvering—that had knocked him out of the first eight. Now, instead of soaring toward Baghdad, he taxied his plane back down into the underground hangar, firing down the Pratt & Whitney engine and popping the canopy. He shimmied down the metal ladder, nodded to the crew techs, and, alone with his thoughts, strode toward the stairway that led up to Operations, where, along with the comm techs, commanders, and generals, he would wait and pray for the return of the pilots. It was the hardest duty he could have drawn.

One hour, three seconds into the mission, the darkening sky lighted up before Ilan Ramon in a phantasmagoria of bright flashes, streaking contrails, and gray puffs of smoke. The skies above the target were pumped full of AAA fire and red tracers streaming in a rising phosphorescent spray from nearly every corner of the shadowland beneath his plane. He had never seen anything like it outside a movie theater. The thin, white contrails, he knew, were probably from SA-7s, shoulder-mounted heatseeking missiles. The larger and deadlier SAM-6s left a wider trail. That was small comfort at the moment. A solid hit in the exhaust from an SA-7 would be every bit as fatal as from a 6. The Iraqi batteries had obviously had enough time to recover from the surprise of Raz and Yadlin's opening assault and warmed up their antiaircraft radars. Triple-A fire now was more directed and intense, a curtain of showering destruction. More SAMs would follow soon. Ramon quickly checked his targeting display, then pushed the nose of the fighter straight down and into the seemingly impenetrable net of 23mm fire crosshatching beneath.

CHAPTER 6

SIXTY SECONDS OVER BAGHDAD

An' if we live, we live to tread on kings;
If die, brave death, when princes die with us!

—WILLIAM SHAKESPEARE,

HENRY IV, PART I

The brown banks of the Gulf of Aqaba rose quickly before the nose of Raz's plane as the ground raced below. The pilots flew in two groups, about seven hundred meters apart, line abreast, and in spread formation. Yadlin flew on the far left, then Raz, then Yaffe and Katz on the right wing. Nachumi's group followed two miles behind. There were several advantages to this formation. If one of the pilots crashed because of fatigue or mechanical failure, he would be far enough apart that he would not hit another plane. The distance between the fighters, broken into two groups, also dispersed the noise of the jet engines, and it had one added advantage: if they were spotted from the ground, it would appear that there were not as many planes in the squadron. The leader Raz held the speed steady at 360 knots, or six nautical miles a minute.

The flight plan took them across the eight-mile-wide gulf, just clipping the southern tip of the Jordanian border, and then diagonally across the northern hump of Saudi Arabia to the western border of Iraq. The Israelis would violate Jordanian airspace south of an abandoned airfield called Haql. Doglegging around the south of the country burned more fuel, but Ivry wanted to avoid as much Jordanian airspace as possible and the country's sophisticated radar network. On the return trip home, assuming there was one, the pilots would fly directly across Jordan from the Iraqi border, but they would be flying at thirty-eight thousand feet and at the speed of sound.

Behind and high above, to his left and to his right, Raz could make out tiny specks that were the F-15s shadowing them. Raz double-checked that his DME (distance-measuring equipment) was in standby mode and his IFF in "standby/receive." He was not due to break radio silence and check in with Command until the 38-degree-longitude point, about one-quarter the distance to Baghdad, another twenty minutes. Then he would break silence only long enough to utter the code word *Moscow*, indicating "so far, so good." Any communication would be in English, the international language of aviation—if overheard, he could easily be mistaken for a commercial flight. As Raz skimmed across the gulf water, close enough to the choppy surface to almost smell the salt spray, he spotted a stunning white yacht at anchor below. Glancing at the fine lines of the ship, he wondered who could own such an incredible vessel, probably Arab royalty or some rich industrialist he thought vaguely, and then, within a second, he blew by the yacht and was across the gulf.

On the deck of the yacht, Jordan's King Hussein, alerted by the distant but unmistakable roar of jet fighters, held a hand to his brow to shade his eyes and peered west toward the Sinai. Several dignitaries and the ship's military officers joined the king on deck, their eyes also straining to the west. To his amazement, Hussein saw what appeared to be four Israeli F-16s streaking toward his ship. As the planes screamed by just overhead, the king could clearly make out the desert-tan camouflage paint on the fuselages and, more alarmingly, two huge bombs hanging from beneath the wings of each aircraft. The wooden planks of the deck beneath his feet trembled as the two fighters thundered by. A moment later a second group of four fighters shook the air again as they missiled by. Immediately Hussein grabbed a secure ship-to-shore telephone and was patched through to Jordanian defense command back in Amman.

"Do you have reports of Israeli aircraft in our airspace?" Hussein asked excitedly.

The king was informed that Jordanian defenses had received no reports of Israeli aircraft.

"Eight Israeli fighter planes just flew over our position in Aqaba, heading east," Hussein informed the colonel on the other end of the line. "They were not more than fifty meters off the ground."

The colonel assured the king that they would investigate immediately. Jordan had signed a nonagression treaty with Israel, and the two nations had not seen any serious skirmishes in years. But the sight of eight heavily armed Israeli fighter planes avoiding radar and heading east was never a comforting prospect. What could they be up to? Hussein wondered.

The king returned to the pleasures of his yacht and his guests who had joined him on this short summer vacation. But he was not nearly as relaxed as he had been.

Amos Yadlin was flying just off the leader's left wing. His eyes were moving back and forth from the HUD to Raz's plane to the contours of the ground below, keeping a lookout for any unmapped power lines or ridges. Already he could see by his fuel gauge the drain caused by the denser air of the gulf and the low altitude. Yadlin could not make out the tiny Arab village of Al Humaydan that he knew from studying the maps had to be somewhere below and to the south. He climbed higher to skirt the first rocky ridges of barren, rust-red mountains, which at some points reached peaks nearly five thousand feet above sea level. Yadlin and the F-16 pilots tightened their formation as Raz led them down the valleys that snaked through the mountain range. Behind, Nachumi's squadron also tightened formation and followed the first team in, with Spector the wingman on Nachumi's left and Shafir and Ramon on his right. Though it was slower going, cutting through the valleys used far less fuel than seesawing up and over mountain peaks, and it made radar detection almost impossible, though the pilots expected Saudi or Jordanian radar—probably both—to spot them eventually.

As he crossed southern Jordan into Saudi Arabia, Katz was amazed at the sight of the desert formations below, a beautiful and eerie rock forest of towering sandstone poles in red, yellow, and orange hues, their long shadows in the late-afternoon sun stretching across the barren floor. The sentinels of rainbow-colored poles appeared to grow out of pyramid-shaped dunes of pure white sand that rose up the trunks. There was not a soul in sight, not a road, not a house, or a tent. He found himself thinking back to the pilots' field trip to Bryce Canyon National Park when they were training at Hill. He had to jerk his mind back to the present: Hey, he reminded himself, this is not a regular flight across the Sinai, this is the real thing!

Traversing the stretch of mountain valleys took less than ten minutes. With Raz in the lead, the planes burst out of the last gorge and soared across the flat, burning desert sands of western

Saudi Arabia, the fabled no-man's-land of the sun's anvil. The desert was nothing but white sand stretching away in all directions, like Katz had always imagined the Sahara. To experience a mechanical failure here and be forced to bail would mean serious trouble. Temperatures on the ground could reach 130 degrees Fahrenheit during the day. The squadron flew on and on, seeing nothing. And then, amazingly, out of nowhere, Katz spotted a lone Bedouin walking in the middle of no-man's-land.

Where in God's name did he come from? Katz wondered. And where could he be going?

After another forty miles, the squadron crossed a narrow asphalt road and a rusted rail line that had once connected the Saudi Arabian city of Tabuk with southern Jordan back in the days of T. E. Lawrence. Raz was following the "blind corridor" in the north that Saguy believed existed between Jordan's radar space and the east-looking Saudi AWAC. Raz heightened his awareness. To the south of Tabuk was a large Saudi air force base. It had wide-ranging radar and occasionally sent out air patrols. As Raz and his squad overflew the road below, the pilots searched up and down the ribbon of asphalt for any signs of traffic on the ground. The road was deserted. Miles and miles of nothing, just sand and sagebrush.

Raz was in a mental zone. He continually checked the HUD, his navigation system, his maps, his wingman, Yadlin. He monitored the rate of fuel use and checked the cockpit computer to determine the most efficient speed and navigation for conservation as the weight of the plane gradually decreased with the fuel burn. Already he was being forced to decrease acceleration to remain at a constant 360 knots with the lighter plane. He made his turns as smoothly as possible, giving the follow-on planes plenty of lead time in order to avoid power spikes that burned more fuel. The baking heat rising from the desert floor began to bounce the aircraft, literally lifting and dropping them in the air currents, making concentration harder and demanding even more mental focus. Though he had switched planes with Yadlin, Raz found himself fighting the same navigation problems he thought he had experienced with his No. 107 on the flight to Etzion on Friday.

Two miles behind, Nachumi followed within visual sighting of Raz's lead group. He was tense, keeping a careful eye on all his instrumentation. The replacement plane felt foreign, and he worried about something going wrong. The plane's handling seemed stiffer. His eye constantly moved from Raz's group to his HUD to his instrumentation to the ground. Oddly, he found himself thinking about his family. He kept seeing mental pictures of his children playing in the yard or sitting at the dinner table. He was aware of a deep longing to be with them.

Back in the command bunker at Etzion, Ivry anxiously awaited word from the first checkpoint. Unlike the U.S. Air Force, where the lead pilot was in command, in the IAF the chief of staff on the ground was in ultimate command of a mission. Behind Ivry on a huge map of the Middle East, his command staff was tracking the progress of the attack group as well as the position of the command search-and-rescue helos and the F-15 support teams. Finally, at 4:23, Avi Sella, crammed into the small copilot seat in one of the Com F-15s, heard a transmission through the bulky, long-range SSB HF (single sideband, high frequency) radio balanced painfully on his lap: "Moscow." One word. He recognized at once the unmistakable voice of his good friend Zeev Raz. Colonel Sella quickly relayed the message and the group's longitude to the 707 circling above Saudi Arabia and then to General Ivry in the Etzion command bunker. Ivry practically flew to the radio when the call came through. Raz and the team were one-quarter of the way there. Ivry reported the progress to Eitan, and the IDF chief of staff immediately phoned Prime Minister Begin and the cabinet ministers, who had gathered anxiously together in Tel Aviv to await the success or failure of the mission.

As the planes continued across the northern Saudi desert to the next check-in position, Point Zebra, Nachumi's mind began wandering again. He found himself thinking of his early days in the Hatzerim flying school, where most of the class had flunked out. He remembered one of the instructors lecturing them: "There are four types of students in flight training. Those who think slow and decide wrong. Those who think slow and decide right. Those who think fast and decide wrong. Those who think fast and decide right. It is the last group we want as pilots. . . ."

The blinking fuel-warning light brought him back to the present. The external wing tanks were nearly empty. Flying just off Nachumi's left wing, Spector also noticed his fuel gauge. He wiped the sweat from his forehead. His eyes were burning, his head pounding. The colonel had awakened that morning with what he had to admit now was a bad case of the flu. He had a low-grade fever and a runny nose. His throat was killing him. Staring into the mirror in the barracks bathroom, he could not believe his poor luck. Spector told no one, however. Nobody ever stayed home from a war because of a cold. After all, he was hardly disabled. He checked his fuel gauge again and waited for the signal to jettison.

Hagai Katz watched his fuel gauge with some concern. Although he had checked with the GD engineers about jettisoning the wing tanks, he still worried. Would the "pans" career into one of the bombs, jamming the bomb's release clips or, worse, detonating it? Or, as when the maneuver was tried with F-4s, would the tanks topple back over the tops of the wings, causing damage to the fuselage, the flaps, God-knows-what? Maybe, Katz found himself thinking, it would be safer just to keep the tanks attached and not risk spiking the entire mission.

Raz was thinking the exact same thing. The moment of truth was upon them, as it were. It was the one part of the painstakingly planned mission that remained more or less an unknown. They could not afford to sacrifice the extra wing tanks they had, so they had not practiced dumping them. No one knew for sure what would happen. He knew the rest of the men were waiting for his cue. But still Raz hesitated. To continue on with the tanks would make the aircraft harder to handle during tracking on final and targeting, there was no question. And, Operations had insisted that the drag from the empty tanks would burn up the crucial amounts of fuel the planes needed to return to home base.

Raz took a deep breath, reached forward, and pulled the switch to release the tanks. He rolled the plane to the left a bit to see if the tanks had cleared the wings. He felt no jolt to the plane as the tanks separated, but did notice an immediate increase in flight speed with the sudden trimming of nearly five hundred pounds of metal.

Ahead of him Katz saw Raz's two external tanks—one, then the other—separate from the wing undercarriages, float in midair for a split second, then tumble end over end to the desert sands below. Katz pulled his switch and felt the same jump in acceleration. Yadlin and Yaffe dutifully dropped their fuel pans next. Raz gave the pilots a thumbs-up to let them know the tanks had fallen cleanly away. Soon the other planes were jettisoning their fuel pans as well, pelting the Saudi desert with sixteen 245-pound wing tanks. They could lie there rusting in the sands for hundreds of years, forgotten markers of a historic mission, Raz thought.

Raz rechecked his INS: less than ten minutes to checkpoint Zebra. He was on course and on schedule. The leader looked left and right, checking his group pilots on both wings—Yadlin, Yaffe, and Katz. Everyone held spread formation, two thousand feet apart. Raz "checked six," looking behind him, and could just make out Nachumi's team: Spector on the left, then Nachumi, Shafir, and Ramon. Everyone was where they should be. He relaxed a little. He thought briefly of the target. Would there be balloons to contend with? How many SAMs?

Nachumi's mind also wandered to the attack—and to the SAM-6s Saguy had warned them about. For ten years he had flown the two-seat F-4 Phantom with a navigator. Though it wasn't much comfort, when going into combat Nachumi, in the back of his mind, always knew that if he were shot down, at least he would have his navigator as company as a POW. In the F-16, however, he was alone. If downed, he would have to face Iraqi soldiers and Iraqi dungeons all by himself. How lonely would he get? Would he crack?

"What are you thinking?" he reproached himself, almost laughing out loud. "I'm a Jew. If they catch me, they'll hang me from the nearest tree."

Out past the left wing, Nachumi saw that the seemingly endless barren stretches of desert were now broken by jagged outcroppings. On the right rose waves of sand dunes, marching south to the horizon as far as one could see. The Saudis' Sakakah airfield should be to the north. He checked his INS. They were right on schedule.

Ilan Ramon was the number-eight man, the last pilot in the group. It was a dangerous position: as the last bomber, he would be exposed to the most AAA and SAM fire, assuming the Iraqi defenders were surprised at all. Either way it was a certainty that by the time he made his run, the antiaircraft batteries would have had time to fire up and begin tracking. Only twenty-six, this was Ramon's first combat mission, and he was tense, no question. He and Relik Shafir had talked about how historic the mission would be, and both had decided they wanted a record of the event. The two pilots made a plan to use the cockpit's HUD TV to record the entire mission. A video camera was mounted in the nose of each plane, complete with audio. But there were only thirty minutes of videotape. Command would want the bombing run on tape in order to confirm target accuracy and assess damage. If the camera were activated at the IP and allowed to run until bombing and escape, it would use an estimated fifteen minutes of tape. That left Ramon and Shafir with a surplus of fifteen minutes to record whatever they wanted for posterity. As he followed Nachumi's lead plane, Ramon kept himself busy by recording various points of interest during the entire journey, so when he returned—*if* he returned—he would have a complete visual diary of the raid.

Up ahead, Raz rechecked his INS. They were at the second checkpoint. He clicked on the radio long enough to utter the word "Zebra," then clicked off. The word crackled in the headsets of the other pilots. Sella picked up the transmission over his headset in the F-15 a hundred miles behind Raz and relayed it on to the 707 and command. Back at the bunker, Ivry heard the second checkoff. The attack group was halfway to target. They would penetrate Iraqi airspace in ten minutes. Ivry grew anxious. The squadron would soon be passing what he considered one of the most hazardous points of the journey—H-3.

An original oil pipeline ran for six hundred miles, all the way from Iraq through Jordan to the Israeli shipping port of Haifa on the Mediterranean. A series of dirt airfields had been constructed at strategic points along the length of the pipeline so that small planes could fly in oil company engineers and techs to conduct maintenance and repairs on hard-to-reach backcountry segments. These airfields were numbered H-1, H-2, H-3, and so on, all the way to Haifa in northern Israel. The H-3 field in western Iraq had originally been a small landing field, but Iraq had years before converted it into a large, modern military base.

Raz's group would pass relatively close to this base on the way to the next checkpoint, close enough, Ivry worried, that the planes could conceivably be picked up by the base's radar—or worse, by an Iraqi MiG on routine patrol. Superstition exacerbated his fears: in the '67 War, Ivry had lost two Mirages during a bombing raid over H-3. As they flew this leg of the journey, Raz's group would observe radio silence and be out of range of the F-15s. Ivry would not know whether the squadron had made it past H-3 safely until Raz checked in at the next point, "Grazen," the name of an Israeli axe.

As the planes grew lighter, Raz increased airspeed to 390 knots. Fuel burn was 75 pounds a minute. They were leaving Saudi Arabia and passing over the border of Iraq. Raz could not tell the difference. The terrain was the same endless miles of barren desert—except that now it was gray and brown. The flight had been uneventful so far, but once in Iraqi airspace, all the pilots increased their vigilance, scanning the pale skies for any sign of MiGs. They were entering a defended area now. Away to the northeast were the Al Habbaniyah and Al Taqqaddum airfields. About a quarter of an hour into Iraqi airspace, the pilots heard the crackle of Raz's voice in their headsets: "Grazen."

Miles behind, Sella picked up the transmission and relayed it on. Inside the bunker, Ivry breathed a private sigh of relief. At least the men were past H-3. He told Eitan, who picked up the secure line and phoned Begin in Tel Aviv: "They're in Iraqi airspace, three-quarters there."

Raz changed his course 30 degrees north, heading straight for Bahr al Milh Lake, west of Osirak. The lake was a critical staging point for the mission: it was the IP, the all-important initial point, sixty miles from the target. There, the pilots would arm ordnance and commence final ingress. It was also the point at which the F-15s would rendezvous before climbing to their predetermined barcaps—that is, predetermined protective overhead combat stations—where they would circle on patrol at twenty-five thousand feet between the Israeli strike force and the Iraqi airfields. Raz and Operations had used a satellite photo (one of the "ill-gotten" KH-11 photos) of the area to fix the IP at the eastern bank of a tiny island in the middle of Bahr al Milh. That would be the final navigation fix from which pop-up, tracking, targeting, everything was calculated. Raz would break radio silence one last time to confirm their position.

At 1734 he spotted the lake up ahead. He was right on time. As he approached from the west, Raz noticed that the lake looked larger than it had on the satellite photo. He began searching for the tiny island so he could update his INS and fix his final navigation. He glanced down. There was nothing. The lake surface was flat and empty as far as he could see in every direction. He was rapidly traversing the length of the lake. He couldn't believe his eyes. His stomach knotted. Where the hell was it? This *had* to be Bahr al Milh. At six and one-half miles a minute, Raz, the mission leader, did not have long to make up his mind. He was soon soaring past the eastern edge of the lake. The roads below matched his military map. There was a town to the left with a tower. There was another village on his right. Ar Rahhaliyah . . . ? That *had* to be Al Mardh. Was he crazy?

It suddenly struck him! It had rained heavily throughout the winter of '80–'81. The lake had obviously swollen with the flooding and the rising rivers. The island was underwater! The waypoint cross on the HUD was sitting some four feet under the muddy waters below.

Abuk! Shit! Raz swore to himself, using the Arab word, since profanity does not exist in Hebrew.

Following on Nachumi's left wing, Col. Iftach Spector was alarmed. Throughout the flight the commander had maintained his own navigational record, tracking times and degrees and checking off map points in a little four-inch pocket-sized book he had whimsically entitled "Paradise Found." Spector saw immediately that Raz had missed the IP. It was a critical miscue. If they were off by as much as one hundred meters, the entire targeting approach would be off. At the IP, Nachumi's team, trailing Raz's squadron at two miles, was to drop back to four miles, or thirty seconds behind, to allow enough time during attack for the concussion and frag pattern of the first bombs to subside before they followed in on target.

As a fallback, the veteran commander Spector had before takeoff selected his own secondary IP: Akhdar Castle, a famed, historical Arabian castle he had read about during his studies, and that he noticed was on the flight path. He had always wanted to see the castle and, since he needed his own backup IP, Spector had factored it into his preflight navigation. Now, as he and the mission team flew above Akhdar, Iftach looked down, satisfying his curiosity as a tourist and student of history, and at the same time verifying their position as yet another enemy invader.

As his group approached the Euphrates River, Spector quickly identified a unique curlicue bend in the river he had noted and circled in ink on the operational map while still at Etzion. Despite the fact that Raz had missed the IP, according to Spector's calculations, the strike force was still on point, some two minutes from pop-up. But the missing IP had thrown Raz a curve, and the squadron leader was clearly doubting his own navigation. But because of radio silence, he could not contact Raz. Spector prayed that Raz would quickly see that he was on track.

Up ahead, Raz rechecked his INS. It showed the squadron to be exactly where he would have assumed they were had he verified the IP. He was right, the IP was simply underwater. But he had overflown the point in any event. Raz reentered his new position and had the computer rescramble the computations. He turned his plane slightly south, some two hundred yards, heading directly for al-Tuwaitha.

On Raz's left wing, Yadlin prepared for the bombing run. He tucked his F-16 in tight to Raz in what was called a "weld wing" formation. He could see Yaffe and Katz maneuvering similarly on the right wing. The pilots would attack in staggered pairs, each pair thirty seconds apart. Yadlin was concerned. He had seen Raz miss the IP and second-guess himself. In fact, Yadlin thought Raz had seemed a bit off the entire flight, but radio silence prevented him from checking in with the leader. Maybe the 129's navigation system was off. Ironically, the 107 aircraft Raz had swapped to Yadlin was flying beautifully. He looked over at Raz on his right and could see his white helmet inside the cockpit. Yadlin felt his hand grow tense on the control stick. This was no time for the leader to hesitate.

Yadlin spotted the ancient Euphrates River up ahead. He was stunned. He had never seen such a huge river in his life. It stretched on and on between its shallow muddy banks, flowing lazily past mudflats and sandpits seemingly for miles. No wonder it had figured so prominently in the early history of mankind. Something on the ground up ahead caught Yadlin's attention. Bizarrely, Iraqi infantrymen on the far bank were waving enthusiastically at the Israelis as they zoomed by overhead, obviously mistaking the planes for their own. What next? He checked to make sure his nose camera was on.

"Now we enter the territory of the bandits," Yadlin said out loud.

The mission planners had calculated that if the F-16s remained below one hundred feet, Iraqi search radar on the ground could not pick them up any earlier than twelve miles out. But after the F-16s crossed the Euphrates, they would be within easy range of Al Habbaniyah and Al Taqqaddum air bases thirty-five miles north. Yadlin anxiously searched the skies to his right and left. They were empty. But for how long?

Katz, too, was anxious now that they were in the populated areas of the Euphrates. They were flying above a rich agricultural region, with neatly plowed fields, electricity lines, roads, and tiny villages, the lights beginning to go on inside homes and stores in the gathering dusk. The Euphrates River looked low inside its banks. The Syrians must be pumping all the water out, Katz thought.

The F-15s that had shadowed the strike force all the way in from Israel began to disperse to their defensive positions. The two F-15s to the north hit afterburners and quickly climbed to twenty thousand feet to form a barcap, the patrol halo, between Raz's squadron and the Iraqi airfields to the north. At the same time the two F-15s to the south burned to twenty-five thousand feet, circling on a patrol barrier between the Israelis and the huge Ubaydah Bin al Jarrah air base in the south. Both patrols turned on their powerful Doppler search radars. With lack of ground "noise" (transmission clutter from both military and civilian communications), spotting the radar bloom of MiGs going wheels-up would be easy for the F-15 navigators. As the two wing groups climbed to high altitude, the F-15 chase planes some ten miles behind Raz's squadron also pulled up to twenty-thousand feet to form a protective umbrella above the Israelis in case Iraq scrambled MiGs from the three airports around Baghdad, including Rasheed and Saddam International. At barcap, the F-15s switched on jamming devices to defeat any SAM radar in the area.

After crossing the Euphrates, Raz pushed back in the cockpit chair to get his butt and back square against the 30-degree slant of the ejection seat. The movement relieved some of the stiffness and got his blood flowing. He also wanted to make sure he was aligned properly in the seat if his plane were crippled by AAA fire. No use surviving if he had his spine snapped in two as his ejection seat shot out of the plane. He swiveled right and left and checked six behind. Everyone was lined up where he was supposed to be. He flipped the switch to arm his bombs and turned on the threat receivers, then activated the VTC video in the nose. His sky was still clear of bandits. Raz was surprised: Operations had predicted that by the time they crossed the Euphrates, Iraqi radar would have picked them up and scrambled MiGs.

He did a quick mental check of the plane: fuel—more than half gone—engine temperature and RPMs—both normal. Electrical systems were positive. He programmed his chaff and flare systems now. During release, as he executed a 7-G turn and climbed at the speed of sound in order to defeat any SAMs chasing his tail, Raz would release his flares and chaff. The flares exploded like gigantic flashbulbs, burning brighter and hotter than the aircraft's exhaust and, hopefully, confusing any heatseeking SAM-6s. The chaff, thousands of strips of confettilike tinfoil, would also draw off radar-guided missiles fooled by the masking of the plane's metal signature.

Raz released the throttle with his left hand, then awkwardly stretched that hand across his chest and used it to grab the control stick on the right side of the cockpit. That allowed the right-handed pilot to use his more dexterous right hand to operate the chaff and flare switches behind him. As he strained backward, keeping his eyes forward, Raz concentrated on the control stick. One slip of his "spastic" left hand and he could nose the plane into the ground.

Dammit, Raz thought, fumbling for the switches behind his seat, it would have been nice if they had designed the cockpit so you didn't have to twist around like a yogi to reach the controls. He had twenty bundles of chaff. He set a computerized program that would eject two bundles of chaff every two seconds during the ten seconds of pop-up, in case a SAM or AAA radar locked onto him during the climb. He programmed the last ten bundles to eject automatically from the tail after he had released his bombs and was thrusting up and away from the target. He turned his attention back to flying. Up ahead he could see lights from a small town.

Khidhir Hamza stood outside a car repair shop just off the dusty main highway from Baghdad, about a mile and a half north of al-

Tuwaitha. It was about 6:30, and the cars normally roaring by during evening rush hour had thinned. The week before, a pickup truck had rammed Hamza's late-model Volkswagen Passat in the side, crumpling his fender. He had brought the car in to the body shop for repairs and painting. A Passat was something of a prestige car in the Middle East, and it was more than worth the money to keep the car looking sharp. As the Atomic Energy project director stood on the lot, waiting for his car, he was startled by a thunderous noise in the west. He looked skyward as a flurry of Israeli fighter planes suddenly shot past the rooftops of the village, two by two by two by two, heading straight for the concrete-and-aluminum dome of Osirak down the road. He could not believe his eyes. Oh my God! Hamza thought. They are going to bomb the Nuclear Research Center. If he had not arranged to have his car repaired that day, he would have been sitting in his office in the administration building this very second.

Yadlin, Yaffe, and Katz glanced down at the tiny village where Hamza stood rooted to the garage parking lot, then prepared for their final approach behind the leader, Raz. Yadlin punched up the targeting control panel on the console above his left knee. The digital screen sprang to life, lighting up in green characters and numbers, showing that the "pickle-button," the red firing pin on the end of his control stick, was selected to fire Sidewinders. With his left thumb he clicked the three-way switch to the right, changing the pickle to the MK-84s. He glanced at the display screen for the readouts on the bombs, checking that they were armed and programmed to drop together at the same instant. The screen display gave him the minimum altitude the bombs had to fall before the delayed fuses would arm: 4.8 seconds. The MK-84s would need to fall at least twenty-five hundred feet before hitting the reactor, otherwise they would fail to arm and be nothing more than duds.

In the lead, Raz had also switched on his weapons and bomb display. The dot of the pipper, the round bull's-eye circle at the end of the vertical bomb-fall line, glowed on the screen. The digital symbol resembled a clock pendulum. As Raz closed on Osirak, the pipper dot moved toward the target icon. When the dot covered the target completely, he would squeeze the "bombs away" button with his right thumb. He checked the INS: eighteen miles to al-Tuwaitha. He watched the mileage click by: fourteen miles, ten miles, six. Up ahead, through the cockpit, Raz could now make out the white, shiny dome of Osirak and the outline of some of the surrounding building. He could see the towering earthen revetments that surrounded the entire compound. They looked mammoth, even at this distance. God, the work that went into all this, flashed through his mind. He squinted, searching for the antiaircraft balloons. There were none. And there was no AAA fire. Raz was puzzled. Maybe they had really surprised them after all. Well, he wasn't going to complain about it.

Behind Blue Flight, Nachumi could also see the dome of Osirak. It was much bigger than he had imagined. The concrete and aluminum dome was covered in mud to reduce its brilliance and help conceal it from enemy eyes. The Iraqis had done that after the Iranian bombing raid. But the dodge was a pipe dream: on the flat river delta, less than a mile from the Tigris to the east, the monumental orb still glinted golden-red in the fading rays of the setting western sun.

Four miles. Time to pop up. Raz pushed back in his seat again. At this point the F-16s were actually flying south, at a 45-degree angle to the target. Raz pushed the throttle to full afterburner and pulled back on the control stick. The Gs pinned him back into his seat as he soared to five thousand feet in four seconds, climbing out of the blinding setting sun behind him. He executed a 90-degree climbing turn and headed straight for the target, then rolled belly-up to maintain positive G forces, keeping the blood pumping to his brain and making it easier to sight the target directly below. While dramatic-looking from the ground, flying upside down for the pilots was nothing particularly spectacular and not at all disorienting.

Raz maintained his focus. In seconds he would roll back and start his dive to the target.

At Etzion, 580 miles to the west, the mission commanders smoked or stared at the radio. It took all of Ivry's willpower not to start pacing. It was 1742 Baghdad time. According to Operations, the attack would, *should,* commence any second—or already had. That was the trouble—no one knew. The strike force was on radio silence. There was no real-time intelligence from Baghdad. They had nothing to do but wait.

Miles behind, circling in an F-15 over Saudi Arabia, Sella sat tensely, balancing the ice chest–sized SSB on his aching knees. Just over the Jordanian border, the line of CH-53 Sikorskys hovered. Near the Saudi border the 707 circled, waiting. In Tel Aviv, six hundred miles west, Begin and his entire cabinet were holed up in the ministry offices, surrounded by a pile of empty teacups and overflowing ashtrays. Waiting.

No one at the ministry was more anxious than Ariel "Arik" Sharon, the notoriously hard-nosed tank commander. He had been a longtime family friend and comrade-in-arms with Yaffe's father. When Avraham had suddenly passed away in 1969, leaving Doobi, not even out of high school, to be the man of the house, Ari Sharon turned up on the Yaffe family's doorstep two days after the funeral. After consoling Mitka and the children, the bulldog-faced general and lifelong rancher took Doobi aside and swore an oath to him: "I will come here every Saturday morning at 7 A.M., and you and I will ride horses together."

Sure enough, for the next three years, until Yaffe was accepted into IAF flying school, every Saturday morning at precisely seven o'clock, the old warrior-statesman would be at the front door, dressed in English riding garb, complete with shined leather boots and riding crop, ready to head out to the nearby stables. The figures of the rotund, genteel general and the beanpole kid beside him would then ride off together on horseback, talking of family, history, and horses, or just nothing at all, except the cool mornings and the clean smell of freshly turned earth. To Yaffe, Arik Sharon revealed a sensitivity and tenderness few could ever guess, and by the time Doobi moved to Hatzerim, the general had become like an uncle to him.

For his part, Sharon felt no less close to Doobi. Now, waiting along with all the other ministers, he worried for all the pilots. But it was especially painful to know that his "adopted" nephew was deep inside enemy territory in harm's way. And there was nothing he could do to help. He tried to keep up a good front for Mitka, but seeing the expression on her face did little to help his own doubts.

Time had stopped. And all Israel, it seemed, waited.

As Raz began to nose his F-16 toward the dome, he quickly double-checked his INS. Something was not right. Osirak was almost directly beneath him already, and he was still on his approach. With rising horror, Raz realized that in the distraction caused by missing the sunken IP, he must have overflown the pop-up point by half a mile. All his calculations had been based on a false point! Now, as he should be beginning his dive, he was too close to the target for his approach on final.

"I am too close!" he yelled out.

Yadlin saw that Raz, instead of beginning a steep 30-degree dive, was still angling straight up. As the leader, Raz was the number one bomber. Yadlin was number two. Yadlin could see the arc of the dome clearly beneath him as he rolled back over. His threat alarm was now ringing loudly. Iraqi radar was finally beginning to lock on. Yadlin's headset was suddenly filled with the jumbled, harried voices from the Iraqi AAA batteries below. The panicked garble grew increasingly intense. The piercing wail of the threat receiver bounced off the glass canopy around him. All hell was breaking loose.

Yadlin knew that he had at most a four-or-five-second window before the Iraqi radar fixed him. He had seen half a squadron lost to SAMs in '73. Nine good men from his unit killed.

The F-16s were already outgunned, and now the leader was hesitating. Yadlin decided in a flash he could not wait.

"I'm not going to end up being hanged in some square in Baghdad because of a screwup," he swore.

Yadlin dropped his nose and cut in beneath Raz's plane, heading straight down "the chute," hurtling 480 knots at the target. He focused complete attention on the HUD, careful to keep the bomb-fall line across the target. He was in a zone: The threat receiver, the engine, the noise of the Iraqi defenders on the ground—everything just faded away. Five thousand feet. The dome raced toward him; the pipper on the display screen creeped ever closer along the bomb line to the target icon. Forty-five hundred feet. Four thousand. He was almost at the predetermined bubble of the frag pattern. The death dot at the end of the bomb-fall line, like a pendulum, crept toward the target icon. Out the glass canopy, Yadlin saw the dome beneath growing larger as he grew closer. He felt the pressure of the tiny red button on top of the control lever. He checked to make sure his wings were level. He had to avoid slipping or the bombs would miss their target.

Thirty-eight hundred feet. The altimeter whirled. Thirty-six hundred. The death dot edged into the target. Thirty-five hundred feet . . . the dot centered, blocking out the dome icon like an eclipse. Yadlin squeezed the pickle and pulled back the control stick. He felt the release clips free the two bombs and the plane seemed to jump ahead with the sudden loss of four thousand pounds. Yadlin quickly clicked the selector button to the Sidewinder fire-control and hit the afterburner, at the same time turning the stick hard left. The F-16 responded immediately, banking radically and climbing as the G-suit bladders ballooned with air, pressing Yadlin's thighs, chest, and head tight against the seat, preventing the blood from rushing to his extremities, which would cause him to lose consciousness. At four Gs, his 165-pound body weighed the equivalent of 660 pounds. He looked over his shoulder back down at the Osirak dome and watched both of his bombs pierce the shell of the cupola and disappear inside, then he was gone like a rocket, racing to high altitude.

As Raz had struggled to modify his approach, he saw Yadlin cut in beneath him. Not a bad move, he thought. He pulled all the way back on the stick, angling the F-16 backward and, finally, all the way over, executing a maneuver pilots called an "overturn," an incredible, circuslike loop-de-loop in which his F-16 turned full circle like a Ferris wheel. Raz came swooping down on the dome at a perfect angle. His eyes were glued to the HUD display, the bomb-fall line tracking toward the target. His right thumb rested on the red button atop the control stick. His threat receiver was ringing in his ears, his headset crackling with Iraqi voices.

The pipper moved slowly into the target icon on the screen. Down, down. Raz reminded himself not to get "target fixation," that was, to continue the dive so close to the ground that he would not have enough altitude left to pull out. At last the death dot completely covered the target symbol. Raz squeezed off the 2,000-pounders and immediately cut ninety degrees left and began his escape. His chaff bundles fired behind him as the thrusters pinned him back, his G-suit bladders filling with air, holding him immobile. Raz switched on his IFF: he did not want the F-15s to mistake his radar blip for an enemy plane.

Suddenly he felt the aircraft shake violently. His heart leaped to his throat. Had he been hit? AAA or SAM? He twisted in his seat but could not make out a thing. The instrumentation showed zero. There was nothing he could do about it now anyway. Raz continued to climb until he rendezvoused with Yadlin. Both pilots began to level off at thirty thousand feet, now far out of range of SAMs and AAA and safe from any pursuing MiGs.

One of the French electricians working at Osirak, Jean François Mascola, stood outside his apartment in the foreigners' compound just down the road from al-Tuwaitha. He heard the fighter planes streaking in from the northwest. Straining to see in the fading light, Mascola could make out a number of planes in the sky above the Nuclear Research Center. He was shocked when the fighters began diving at the Osirak dome. Though not partial to Saddam Hussein, he had worked long at the nuclear reactor and made friends with many of the Iraqi technicians and scientists. He immediately worried for their safety.

To Mascola, the planes diving at the Osirak dome looked like something out of a movie. Flames leaped into the evening sky and the ground shook with the explosions of the powerful bombs. But unlike the world of make-believe, he heard no sound of AAA fire or saw no streaking tracers in the heavens for a long while. Then, finally, after the first detonations, the sky erupted in a fireworks display of missiles and AAA. It was somehow both beautiful and awful to behold.

Just seconds behind the group leaders, Yaffe finished his pop-up and roll and began his approach on final. Katz followed close behind him. Neither pilot had seen Yadlin cut in front of Raz and release first. Yaffe felt the adrenaline flowing. His muscles tensed. Over his headset he thought he heard pilot chatter between the Iraqi Tupolev fighters stationed at Al Habbaniyah to the northwest. How did the Americans put it? The shit would hit the fan soon. The ground was now dark beneath him and it was becoming difficult to distinguish the horizon line from the darkening sky. Suddenly the ground below seemed to jump at them. Sparks and flares and tracers exploded all around them as Katz began to zigzag in order to become a harder target to hit. Finally, up ahead, he could see the Osirak dome caught in the final rays of sunlight. As he neared, Katz could see that the dome had already partially collapsed, its shiny arc marred by jagged holes left by Raz's and Yadlin's bombs.

I've trained my whole life for this mission, Katz thought. I have one chance to do it right. Don't screw it up!

As he watched the pipper moving toward the target, Yaffe, diving just ahead of Katz, thought he saw white puffs out of the corner of his eye to the left. Soon the sky was filling with them. Those are not clouds, Yaffe realized with horror. They're shooting at us! Sealed within the cockpit canopy, drowned out by the whine of the Pratt & Whitney and the ringing of the threat receiver, Yaffe could not hear a thing going on outside. But he knew, for whatever reason, that the mysterious absence of antiaircraft and SAM fire was over. Nothing to do about it now, he told himself. He focused on the bomb-fall line. The crippled dome rushed toward him. The delayed fusing on Raz's and Amos's bombs had kept them from exploding so far. He had a clear shot—4,000 feet, 3,700, 3,500. Now! Yaffe pulled back on the control stick and at the same time pressed the red button, pickling off his two bombs. They fell cleanly away without a hitch. Seconds later Katz released his MK-84s. A total of eight 2,000-pound bombs had crashed through the now-gaping reactor dome. Yaffe and Katz climbed to altitude, their chaff bundles igniting behind them. Raz's Blue Flight was away.

Nachumi's group began its final run, he and Spector in the lead. Nachumi had navigated perfectly—the two lead fighters were exactly thirty seconds behind the number three and four bombers, Yaffe and Katz. The sky above al-Tuwaitha was filled with streaking tracers and gray clouds of exploded AAA. An air-to-air fighter pilot, Nachumi was familiar with one-on-one encounters. But this was more antiaircraft fire than he had ever seen in his life. Ten years later he would be reminded of the scene again while watching CNN's images of the skies above Baghdad during the Persian Gulf War. The F-16's threat system still showed no SAM batteries lighted up. But it was literally raining AAA. And it was scary as hell.

Worried about Shafir and Ramon, who would be even more exposed, the second team leader closed the distance between Yaffe and Katz and his plane. Spector, in weld-wing formation, closed along with him. Such in-flight proximity was highly dangerous, but Nachumi decided protecting his seven and eight bombers was worth the risk. "Well, what are you going to do?" Nachumi mumbled to himself.

He looked out the canopy and saw Spector just off his left wing, exactly in position. His eyes switched from the HUD to the dome and back again. He had seen Yaffe and Katz drop their ordnance, but still there was no dust, no smoke. Had they hit the dome wrong? he thought anxiously. Then he remembered that the dome would contain much of the force of the initial explosions. And, of course, the first bombs were on a delayed fuse. Nachumi drove them down, past 3,500 feet. He wanted to be absolutely sure to hit the target. At 3,400 feet he pickled off his bombs, then angled ninety degrees left and began his climb. As his afterburner kicked in, he felt the plane being buffeted about. What the heck? he thought. Was it a SAM? He scanned his instruments. Nothing.

"Let's get out of here!" he yelled into his radio mouthpiece, not intending the message for anyone in particular.

Spector began his final approach just seconds behind Nachumi. He had been fighting the flu and fever the entire flight. He willed himself to ignore it, but the fever was sapping his strength nonetheless. Spector initiated pop-up, shooting to five thousand feet in seconds. The blood seemed to drain from his head. As he nosed down he felt dizzy, light-headed—and then for a brief moment, he lost track of where he was. He shook it off. Had he blacked out? He wasn't sure. The colonel continued his dive, struggling to keep his focus on the pipper and the bomb-fall line while at the same time keeping a wary eye out for MiGs. He realized that his lack of training time in the F-16 was a handicap at the moment. The altimeter whirled before him. He followed Nachumi down. But something was wrong. He was not where he was supposed to be. The pipper was not dead on the target and he was at 3,500. At 3,400 feet he pickled off the bombs. As the MK-84s disappeared beneath him, Spector hit the afterburner and shot skyward. Behind him, Osirak quickly fell away, shrinking into a tiny square on the earth below. But it was no good. Spector was far too professional, far too good a combat pilot not to know. He had missed. His bombs had not hit the target. He had failed. The realization came to him like a knife to the gut.

Shafir and Ramon, the final pair, followed immediately behind Spector. They did not see that their commander had missed the target. The sky was too thick with AAA and the unmistakable contrails of shoulder-mounted, Soviet-made SA-7 missiles. Smaller than the radar-controlled SAM-6s, the sleek, heatseeking SA-7s were every bit as deadly. Shafir breathed hard through clenched teeth, the threat receiver a background of dissonant noise. Deadly puffs of clouds continued to mushroom all around him. Ramon spotted an SA-7 streaking by on the right not more than twenty meters from his wing. The sight of the deadly contrail made him uncomfortable, but he was not surprised. All their briefings had told them to expect as much.

Shafir and Ramon continued their approach on final. And then suddenly, the dome beneath them exploded outward in a mammoth cloud of black, acrid smoke followed by spectacular flames leaping hundreds of feet into the air. The delayed fuses dropped by Raz and Yadlin had detonated. Shafir and Ramon were now forced to release blindly into the smoke, hopefully avoiding the shrapnel and debris of the immense conflagration below. Well, Shafir told himself, the number seven and eight spots are always the riskiest; it was the lot they had drawn.

At 3,400 feet Shafir pulled up, followed in sync by Ramon on his right wing. The two pilots released their bombs into the crumbling dome, then rocketed up and away to the running rendezvous with the rest of the squadron at 30,000 feet. As Ramon climbed, the last pilot out, the dome of Osirak erupted below in one final thunderous fireball. Seen in the videotape later, the exploding dome would look very much like a special-effects scene. For the first time in history, a nuclear reactor had been bombed and obliterated.

All that remained was to return home.

Jacques Rimbaud, another French technician, sat enjoying a Pernod and water on a patio café in the small village next to the al-Tuwaitha plant. Like most of the French techs, Rimbaud took Sundays off, especially Sunday evenings, even though it was a workday for Iraqis. He was jolted from his reveries by the deafening scream of jet fighters, which soared overhead seemingly out of nowhere. As Rimbaud watched, it looked to him as if two planes made a pass over the reactor, and then a second pair dropped bombs. These planes, he excitedly told the other patrons who had rushed outside to see what the commotion was about, were followed by four fighters.

"They are taking pictures of the area to confirm damage," he explained, taking for granted that the fact that he had seen the planes first made him the tacit expert on the scene.

As the planes circled and raced west, Rimbaud ran down the dusty street to the Nuclear Research Center to see if his office was damaged. The guards at the main gate, still stunned from the attack, would not allow him in at first. Finally, after a good deal of yelling and explaining, and after checking his worker identification, the guards allowed him inside. Within the walls of al-Tuwaitha, soldiers and firefighters were rushing toward the reactor and the administration building. Individuals were running in all directions, some carrying files and paperwork, others seemingly confused. Security men with bullhorns were shouting orders to crowds of workers and other soldiers who did not bother to listen. An acrid smell of explosives filled the air. Huge flames leaped from the dome in the center of the grounds, casting Stygian-like shadows across the sweaty faces of the fire crews and the panicked security guards, already afraid of repercussions.

Rimbaud was stunned. He saw immediately that the damage to Osirak had to have been done by more than bombs from two planes. The reactor was demolished. He would not have believed that concrete and steel could be so smashed and twisted—like a child's toy. A fellow worker recognized Rimbaud and approached him, clearly upset.

"All these bombs must have fallen within one meter of the target!" the man exclaimed, astonished.

The lieutenant checked his watch, then looked at the members of his CSAR crew. The pilot and the navigator up front held the chopper steady, listening intently for any comm over the radio. The pilot pulled back on the stick, steadying the CH-53 and fighting another updraft of warm desert air. They had been holding position just over the Jordanian border for nearly four hours, keeping a sharp eye out for bandits and waiting intently for the first crackle from one of the pilots' PRCs. The lieutenant saw the fatigue in his men's faces, the slightly swollen eyes, the expressions. There was no chatter, no complaining, the hallowed right of the noncom in any army. They were too tired to bitch. There had been no word from command and nothing from the Israeli aircraft. That was good news. But it was hard, waiting and wondering, crammed into the claustrophobic hindquarters of the helicopter, fighting for space with M-16s, ammo boxes, first-aid kits, infrared binoculars, field radios, and other sundry tools of the trade of search and rescue. Finally, as the sky to the east began to darken, the radio came to life. It was from Beersheva, giving the team the "all clear" signal to return to base. The men shrugged, relieved that they could return home to hot food and hot showers. But still, there was a nagging sense of incompleteness. For not one of them knew what they had been waiting for or whether or not they should be celebrating or grieving.

CHAPTER 7

CHECK SIX

Home is the sailor, home from the sea,
And the hunter home from the hill.

—ALFRED, LORD TENNYSON

Khidhir Hamza stood frozen in front of the auto repair garage as the rumble of explosions came from the direction of al-Tuwaitha. In between, Hamza could make out the faint staccato of AAA fire. One final immense blast shook the ground beneath his feet. Seconds later another thunderous clap rolled over the garage as the eight F-16s, high in the sky now, flew back the way they had come. And then there was silence. As if in a dream, as quickly as they had appeared, the fighters disappeared, vanished into the west like ghost ships.

Shaken, Hamza watched a plume of gray-white smoke snake into the sky to the south. He knew immediately what had happened. And he knew that the damage at the al-Tuwaitha "campus" would be nothing like the Iranian attack nine months earlier. He felt queasy.

The wail of an ambulance raced down the highway from the direction of Baghdad. As the emergency vehicle went speeding past, lights flashing, Hamza recognized a colleague from the Nuclear

Center driving in the opposite direction, from al-Tuwaitha. He waved him over to the side of the road.

"It's the reactor," the man told Hamza, his eyes wide, his forehead beaded with sweat. "They destroyed it. It's the Israelis this time."

Hamza did not doubt it. Who else had such aircraft? He had seen that the planes streaking above were a newer, highly sophisticated fighter. Certainly American-made. For some time the al-Tuwaitha scientists had been concerned about the Israelis. They knew that Mossad and the IDF could be ruthless when threatened. Rumors about the death of al-Meshad and the La Seyne-sur-Mer explosions had spread throughout the center. Like the rest of the Osirak engineers, Hamza for some time had felt that no one was safe from the reach of Israel's wrath. But this strike marked a radical shift in the Jewish nation's campaign against Saddam.

Hamza retrieved his VW Passat and drove home as fast as he could. Curiously, there was nothing on Baghdad radio about the air raid. As he pulled into his driveway, Hamza saw his wife waiting for him at the door. She was nearly hysterical. The television had gone out completely, the screen filled with grainy snow and interference. And then a neighbor had called to tell her that planes were attacking Baghdad. Hamza knew the television interference was from electronic jamming. The mission was well planned—and obviously Israeli. He calmed his wife down, assuring her that they were safe. Israel had already accomplished its objective.

He poured himself a scotch and then called the deputy director at Atomic Energy.

"The reactor is gone," the deputy director told him. "But Isis is okay. And they missed the Italian labs." Small consolation, Hamza thought, but he did not share his opinion with the bureaucrat.

At thirty-five thousand feet, twenty miles west of Hamza, Raz focused on gathering the pilots together for the return home. And still he fretted about the buffeting of his plane. He was sure he must have been hit. He signaled Yadlin to close in and give his plane a visual inspection.

Yadlin dropped below and scanned for any signs of damage to the fuselage or wings, any bullet holes or frag hits. He found nothing. Pulling even with Raz's fighter, he signaled a thumbs-up: everything looked okay. Raz was puzzled. Well, he thought, he would have to wait until they landed to review the video and figure out what the heck had happened. He opened his radio frequency, waiting for the pilots to check in. Raz and Yadlin were in visual contact: Blue One and Blue Two. Yaffe radioed "Blue Three," dropping in on the other side of Raz's wing. Then Katz, "Blue Four." Nachumi's team was next. After what seemed like an eternity, Nachumi finally radioed in, "Blue Five." He was followed quickly by Spector: "Blue Six." Shafir checked in, "Blue Seven." Ramon was next. But Raz heard nothing.

Raz waited.

Where the hell is Ilan? Raz thought, growing concerned. He felt a sudden hollowness in the pit of his stomach. Ramon had had the worst of it as the number eight. Did he make it?

Raz twisted in his cockpit, anxiously scanning the skies around him. Dammit, the youngest . . . why would we give him the hardest position? What were we thinking? Where in God's name is he?

Raz fingered the radio switch, then searched the skies again. He could stand it no longer. Against orders he switched on the radio and broke silence.

"Blue Eight, Blue Eight. Check in! Where are you?"

Following far to the rear, Ramon was startled to hear his call sign break in over the radio headset. He realized that in all the excitement, he had forgotten to check back. *"Shi'in"* ("Shit"), he swore to himself in Arabic.

"Blue Eight, roger," he radioed. "Joining up now."

Raz blew a stream of air between his teeth. Thank God. He switched on his radio again, then sent out the code word to Sella for "all clear." *"Charlie,"* Raz radioed.

Circling high above Saudi Arabia, Sella heard the long-awaited code word. Everyone was accounted for. Everyone was safe. They had survived. The mission was a success. He felt elated. But there was no time to celebrate. It was not over yet. Sella radioed back to Etzion: "Charlie!" Mission completed. All the planes were returning to base.

The command bunker broke into a spontaneous cheer at the communication. What had seemed an eternity, during which the mind conjured the most horrendous images, had ended with a word. The pilots were safe. Ivry nearly trembled with relief. But he quickly collected himself. They were not home yet. They still had to cross Jordan. He passed the word on to Eitan. The strike force was heading home.

The chief of staff called Begin at his office in Tel Aviv.

"Mr. Prime Minister," Eitan said, "the mission was accomplished without losses. The planes are returning to base."

"Let me know when they get back," Begin replied, trying to tamp down his excitement.

When he hung up, despite his best efforts to keep his emotions in check, the prime minister burst out with the news to his ministers.

"The pilots are home safe! There were no casualties. The mission has been a success!" he all but screamed at the assembled ministers, their faces twisted with a mixture of dread and anticipation.

The cabinet room fairly vibrated with the sudden release of

pent-up tension. A wave of intoxicating relief broke across the prime minister's residence where they had gathered. Cabinet ministers, grinning widely, some with tears in their eyes, slapped one another's backs or hugged one another. "Praise God," many uttered. Yaffe's mother, Mitka, tearing up herself, moved quickly to find a telephone. She had a call to make. Hurriedly, she dialed her son's phone number at the base residence, her hand nearly shaking as she listened to the phone ring on the other end.

"Hello," Michal Yaffe answered, tentatively.

Doobi's wife had been wandering from the living room to the kitchen and back for hours, sitting and flipping through a magazine without seeing the pages, then tossing the magazine down and walking back to the kitchen to stare out the small rectangular window that looked onto the narrow street outside. Anything not to stare at the telephone. On the table next to the sink was a bottle of unopened French cognac and two empty glasses. Ramat David had been quiet for hours now, the distant sound of her husband's F-16s nonexistent all afternoon.

Suddenly, just after seven o'clock, the phone rang. Her heart seemed to clutch. Her lungs felt like they were collapsing. Now she understood the familiar saying about your heart leaping into your throat. Except they never mentioned the wave of nausea that hit the stomach at the same time.

On the other end of the line, Michal recognized her mother-in-law's voice. My God, what would she tell her. . . . ?

"Pour yourself a brandy," Mitka said simply, using the code the two women had agreed upon: the signal that Doobi was back safe. And alive. Michal felt her knees starting to buckle, literally. Her strength just seemed to gush out of her.

"Oh! Yes. Of course," Michal exclaimed. "Thank you, Mitka. I will! Thank you."

The F-15 pilots, circling in their barcaps, also heard the code word "Charlie." Immediately the pilots climbed to altitude and fell in formation behind the F-16s, ready to supply cover if needed. Their search radar continued to show no MiGs in the area. The raid had caught the Iraqis by complete surprise.

Raz climbed to 38,000 feet, where the intelligence forecaster had predicted the weakest headwinds. He checked his computer readout: indeed, there were virtually no air currents. But at this altitude, the F-16s' exhaust would leave behind contrails in the cold, moist air, making the planes easier to spot. The squadron still had to cross Jordanian airspace, where MiGs could well be waiting for them. Raz zoomed up to 40,000 feet. At that elevation there was less moisture, and the planes would not produce telltale contrails. But his HUD showed headwinds there at 125 knots. That would cancel out most of the fuel efficiency they would gain by flying in the thinner air. Should he take the safer, slower route, flying virtually invisible at 40,000 but risk running out of fuel, or should he head for home as fast as possible at the lower, more fuel-efficient altitude? Raz opted for the faster route and descended back to 38,000 feet. If they ran into MiGs . . . ? Well, they would deal with that if it happened. He set a direct course between Baghdad and Etzion, cutting straight across the northern "panhandle" of Saudi Arabia and the southern portion of Jordan. The eight fighters roared west, streaking bright white chalk marks across the blue skies behind them.

Seeing the F-16s level off at 38,000 feet, the six F-15s climbed to 41,000 feet and mothered them home.

Raz was so relieved to be finally heading home that he found himself singing in the cockpit. He smiled, a little embarrassed. The Jordanians could still challenge them. He kept an eye on his instruments as well. He continued to suspect that his plane was experiencing mechanical difficulties. Behind him, Nachumi and Spector were also concerned about the violent shaking their planes had experienced during their escape maneuvers. The two pilots flew above and below each other, inspecting for any external signs of damage. Neither could detect anything on the other's fighter. Even so, Nachumi continued to monitor his instrumentation carefully.

Meanwhile, up ahead, Yaffe anxiously eyed his fuel indicator. Since not topping off, he had worried that he did not have the fuel to get home. Ramon couldn't help himself. He broke radio silence just long enough to relay a one-word message to Yaffe: "Alhambra!" It was the name of Tel Aviv's finest restaurant—the payoff he had wagered with Yaffe that they would all make it back safe. Doobi was on the hook for the most expensive dinner of his life.

After an hour's flying time, the strike force crossed into Jordanian airspace just miles north of the country's southeastern corner—far from the radar installations and airfields around Amman. Like a horse nearing the barn, the pilots now raced toward Israel, at one point shaking the ground below with a thunderous sonic boom as they shot homeward at the speed of sound. Behind them, north of Amman, IAF radar picked up MiG chase planes scrambling west and south. But Raz's group was too high, too fast, and too far ahead to be engaged.

General Ivry picked up the radio microphone and, for the first time anyone at IAF command could recall, directly called the mission pilots from the command bunker.

"Good job!" he said. "Now make a safe landing."

Raz racked his brain, desperately trying to think of something clever or memorable to say. He was not the talkative type. Shafir and Yaffe were the phrasemakers. All he could manage was, "Roger." But that did not mean he could not feel the weight of the moment. After ninety minutes he could finally make out the lights of Eilat on the west banks of Aqaba. He felt something akin to an electric jolt, then began his descent from 38,000 feet.

Flying behind, off the right wing, Katz followed in his descent. For Hagai, the entire return flight had been surreal. The flight path west was directly into the red glow of the setting sun. The F-16s were flying just a tad slower than the sun was setting. Instead of the usual four-minute sunset, Katz, at 38,000 feet above the curve of the earth, had watched the sun in front of him set continually for forty minutes. Now, as the planes dropped down toward the earth, they dove out of sunlight into darkness in seconds. In a heartbeat, all was nighttime. It was an eerie feeling.

Yaffe was especially relieved to be landing. He had been second-guessing his decision to take off without topping off and worried about running out of fuel for the last half hour. But the lack of headwinds saved the squadron many gallons, and his gamble paid off. The landing lights at Etzion were on; the rest of the base was blacked out. The soldiers on the base did not know the details of the F-16s' mission, but they knew that something important had happened. After the planes had taken off, standby fighters had taxied to the end of the runway, ready to go wheels-up. The maintenance crew had busied itself in various ways, knowing the mission would be a long one judging from the amount of fuel the jets were carrying. Most of the crew chiefs played cards or dominoes in the hangars, but by the time the planes were back in Israeli airspace, the majority of the men were out on the tarmac or waiting anxiously outside the maintenance hangars, looking to the darkening east.

Finally, just after 1900, they could make out the first dots in the sky. They anxiously counted the numbers. One, two, three . . . six . . . finally, eight! They were all back. The F-15s broke off and headed directly for Tel-Nof in the north. For ten minutes the eight planes dropped one by one to the tarmac, their tires smoking rubber as they hit the runway. All accounted for. The crew was ecstatic.

In the bunker, Eitan called Begin for the last time.

"All planes have returned safely," he informed the prime minister.

"Barach hashem!" Begin sighed.

"Blessed be God."

He signaled Mitka to retrieve the bottle of vintage French cognac he had promised himself half a year ago that he would drink to toast the success of the mission. Mitka set out glasses for the entire cabinet. Everyone suddenly felt thirsty.

Five hundred and eighty miles to the east, al-Tuwaitha was in chaos. The Nuclear Research Center security guards and Iraqi military troops carrying Kalashnikovs raced back and forth across the compound, challenging anyone not in uniform and dodging fire engines and emergency vehicles. Everyone was jumpy. Darkness had fallen, adding to the confusion. The shadows were sliced by the headlights of fire trucks and jeeps racing across the grounds. Eight people had already been confirmed killed. One was a Frenchman named Damen Chaussepied, a twenty-five-year-old nuclear technician caught in a lab hallway attached to the reactor. Osirak itself was completely demolished, the once monumental dome looking like a broken eggshell in the flickering flames, which still danced high into the sky from deep within the reactor.

As Mossad had discovered, the Iraqi units manning the AAA batteries at al-Tuwaitha had made it a habit to break for supper every night before six o'clock—one of the reasons Ivry had planned the attack time for after 6:30, Baghdad time. Before heading to the center's cafeteria, the battery teams inexplicably shut down their ZSU-23–4 and SAM radars. The men were just sitting down to dinner at the cafeteria tables when they heard the first bombs hit the Osirak dome. Surprised, they raced back to their stations. But by the time the gunners reached their AAA emplacements, Raz's team, the first four planes, had already dropped their bombs. As Yaffe dived, the antiaircraft gunners were desperately fumbling to start the radars, which were completely cold. Most of the ZSU radars had nowhere near enough time to warm up in the four minutes the planes were in range. The gunners decided to begin targeting manually, without the aid of radar or computers. The fire was wild and intensive. In the excitement, AAA gunners manually tracking the diving F-16s dropped their line of fire so close to the ground that they actually began mowing down their own gun emplacements on the opposite side of the compound, killing several soldiers and wounding at least a dozen men. Meanwhile, the explosions from the bombs sent debris sailing high into the air, crashing back down to the ground and wounding workers and guards scurrying across the compound. The cries of the wounded and the frightened could be heard everywhere across the installation.

As darkness fell at Etzion, one hour behind Baghdad, Rani Falk waited along with the crew chiefs, who signaled the taxiing F-16s in with their red-tipped flashlights. The base was still dark, radio silence continued. The planes were directed to their underground hangar glowing bright under the glare of floodlights. The pilots emerged from their planes, blinking momentarily in the harsh glare after spending hours in the darkened cockpits. As they climbed down the ladders, the ground crews surrounded them, patting their backs and congratulating them. The fliers' faces were flushed with the mix of emotion and adrenaline.

Raz, Spector, and Nachumi all carefully checked their planes, examining the wings and fuselage for any signs of flak damage.

They told their crew chiefs to look over the fighters inch by inch to make sure there were no bullet holes. Then they joined the rest of the men to be transported to the debriefing room.

Iftach Spector stood to the side, visibly upset.

Nachumi was pretty sure Spector had missed the target. The other pilots, too, had immediately sensed that something was wrong. They began to approach their longtime commander.

Nachumi clasped him on the shoulder.

"We did enough damage to the reactor," he said, hoping to console him.

"I want to see my video!" Spector snapped, waving the men away.

It was clear he wanted everyone to know something had gone wrong. It was equally clear he wanted to be alone. Katz was surprised. He had never seen Spector so rattled.

The pilots walked straight to Operations. Eitan and Ivry had already flown on ahead to Tel Aviv to brief the prime minister. The squadron was served coffee and sodas. The men unwound fast, an easy casualness quickly replacing the frayed edges of tension and nerves they had lived with since taking off seven hours earlier. They discussed the mission, the attack, the surprising lack of AAA until the final minutes. Nothing was said about Raz missing the navigation. Neither Raz nor Yadlin mentioned that Yadlin had cut in beneath Raz, or the fact that the first two bombs had been dropped by the number two flier. Unmentioned also was Raz's astounding 360-degree backflip over the target, under enemy fire, and then dropping both two-thousand-pounders with 100 percent accuracy.

At last the crew chief brought in the nose-camera videos. The men quickly took their places around the TV monitor. It was what they had been waiting for. Everyone was eager to see his video. The videos showed the view of the gunsight camera, the corners of the frame delineated by brackets. In the middle of the screen was a cross indicating the location of the pipper. On the lower left side were digital readouts of the plane's airspeed, navigation, and other vital signs. The men watched each pilot's video intently, grading the quality of the attacks—good passes or bad passes. Shafir's and Ramon's cameras had documented the entire mission. As Relik's camera recorded his plane crossing the Euphrates, Iraqi soldiers could be seen waving at him from the far bank. The pilots laughed.

Raz and Yadlin's videos were shown first, the dome of Osirak rising clearly in the middle of the huge compound, the HUD gunsight cross flitting across the frame like a butterfly. No AAA fire could be seen. The dome of the reactor raced toward the camera, and then the bombs could be seen piercing the shell, leaving behind a gaping crack. Then came Yaffe and Katz. As their ordnance fell, the dome crumbled inward, leaving a jagged open mouth. By the time Nachumi's video screened, the crown of the decapitated dome looked like a softboiled dipping egg. As Ramon, the final bomber, sighted the target, the videotape showed the cupola below spewing huge funnels of black smoke from deep within the reactor. Ramon's two bombs could be seen disappearing into the smoke. But as the plane began to climb, the nose camera captured the Osirak dome below exploding outward, erupting in a volcanic conflagration of flame, utterly demolished.

At the sight of the final destruction of Osirak, the pilots broke out in spontaneous whoops. Their mission had been a success: the complete annihilation of Israel's most deadly threat. Throughout it all, Spector stood off to the side, watching silently. He obviously felt terrible. For the videos had made clear what Spector himself had realized alone in his cockpit: the commander had missed the target. With both bombs. Everyone else had targeted with 100 percent accuracy.

Outside, Spector was disgusted, but he was determined not to show weakness. He was a leader. He did not have the luxury of being vulnerable. He also made up his mind that he was not going to use the flu as an excuse.

"I missed," he said, with forced equanimity. "Something happened. I'm not sure what. . . ." He shrugged. "But there is no excuse. Thank goodness you were there to back me up."

No one dared ask him what had happened. If he wanted to tell them someday, he would. As the pilots turned toward the runway, Raz smiled grimly to himself and shook his head.

"He was punished!" he thought. "He went over everyone's head, and he was punished!"

Nachumi was extremely upset by Spector's obvious pain. By pulling in so close to the first four planes, had he taken attention away from his wingman at the most critical part of the bombing, when acquiring the target? The thought would haunt him for the next twenty years.

The squadron made ready to leave. The long day was not over yet. They still had to return to Ramat David and then fly in small planes to Tel Aviv to meet with the command and support teams who had gathered to congratulate the pilots and hear firsthand the details of the mission. Out in the maintenance bays the crew chiefs were busy refueling and cleaning the F-16s for the return flight north. The maintenance techs had found nothing wrong with either Raz's or Nachumi's plane. The only logical explanation was that they had dived so low to release their bombs that the shock waves from the explosions had shaken the planes before they could escape.

The men zipped up their flight suits, grabbed their gloves and helmets, and climbed the metal ladders back into the cockpits. Takeoff was much simpler and quicker than it had been some six hours earlier, when the planes were overloaded with fuel and bombs. Raz led the team home on full thrusters, streaking at supersonic speed the entire trip and rattling the windows of the towns and kibbutzes below—and breaking IAF rules against flying at supersonic speeds over civilian territory. He was sure they wouldn't object this one time. The men were back at Ramat David in less than half an hour.

After landing, each of the pilots drove to his on-base home to clean up and change clothes before leaving for the assembly in Tel Aviv. The pilots were still sworn to secrecy, but many now elected to tell their wives what they had been up to for the last eighteen months. Yadlin's and Yaffe's wives, who knew about the raid in advance, were nearly beside themselves with relief after enduring eight hours of hell. Katz told his wife for the first time. She was stunned, but the two shared a laugh over her unconscious but nonetheless uncanny prognostication at his departure. The pilots showered and shaved and changed clothes, and, an hour later, were back out on the runway for the thirty-minute hop down to Tel Aviv on a small, cramped, twin-engine prop plane. As the plane flew south in the darkness, Yaffe shook his head in his seat. After all the meticulous planning and expense of the last year, this was a pretty bush league way to end the mission: to have all eight of Israel's new heroes shoved into a rickety puddle-jumper that could crash at the bat of an eyelid.

The plane landed at the tiny airfield in north Tel Aviv. A van quickly transported the pilots to IAF headquarters just outside the city, where they were dropped off in front of a small auditorium. Heavily armed security ringed the building. As they marched down the aisle of the auditorium, the men were greeted by a standing ovation from Ivry, Eitan, the senior command, and the dozens of operational staff and IAF support teams that had toiled in secret over the mission planning for the last year and a half.

Again, the men were peppered with questions about every detail of the mission, and especially the attack. The excitement was such that soon the staffers were shouting over one another to ask questions. Raz had a hard time deciphering most of them. Instead he continued smiling and repeating: "It all went according to plan. There is nothing to say."

Not satisfied with Raz's nonanswers, the inquisition turned to the other pilots, relentlessly pressing for every detail.

"It was just as we planned in the briefing," Yaffe answered. He recited a quick rundown of the mission. There was some AAA, maybe SA-7s. The defenders had been surprised as there were no MiGs. They had caught the Iraqis with their pants down.

Ivry and Eitan were nearly bursting with frustration. This was probably the most historic event in the history of Israel, and the pilots were treating it like some boring job!

"Nothing to write home about," Katz repeated.

Relik Shafir was at his nonplussed, self-effacing best.

"I was just at the right place at the right time," he quipped, shrugging off the compliments. "It's like the game of golf—the more you practice, the luckier you get."

Undaunted by the pilots' disappointing, matter-of-fact answers, the Operations officers and ministers began debating the mission among themselves, examining and rehashing the minutest detail of every action.

Finally, Avi Sella spoke up.

"What was the meaning of the code word *Alhambra*? I did not see that in the operational notes. We worried we had missed something."

Raz smiled. "Why don't you explain that, Doobi."

The room went quiet, waiting for this momentous intelligence.

Looking sheepish, Yaffe told the room about the bet he and

Ramon had made the night before the attack. Ilan was only reminding Yaffe that he owed him a dinner at the Alhambra Restaurant. The crowd broke into a cheer. Ivry and Eitan both demanded invitations. Finally, toward midnight, almost punch-drunk from exhaustion and spent adrenaline, the eight pilots climbed aboard the prop plane for the final trip home to Ramat David— and a few days of rest. The next day, Monday, was a national holiday, Shavuot, a celebration of the Feast of the Pentecost, the giving of the law to Moses and the Jewish people by God. It was one of the most popular holidays of the year, a summer festival that combined the feel of the United States' Fourth of July and Halloween, a day of picnics, concerts, beach cookouts, and hayrides.

The holiday was one of Katz's favorites. As he made ready for sleep that night, he thought for a moment how strange it was that less than twelve hours ago he had wondered briefly if he would ever celebrate that holiday again.

Richard V. Allen, President Reagan's head of the National Security Administration, was at home outside Alexandria, Virginia, Sunday afternoon, relaxing on the sundeck, drinking iced tea and flipping through the weekend homework of position papers, memos, and classified reports when the telephone rang. A White House aide in the Situation Room, the round-the-clock communications center located in the basement of the West Wing, reported that Israel had just informed the State Department that its air force had bombed the Iraqi nuclear plant at al-Tuwaitha.

"When?" Allen asked.

"About five-thirty, six o'clock their time."

Allen quickly rang off and was patched through to President Reagan, who was spending the weekend at his favorite retreat,

Camp David, deep in the Catoctin Mountains of Maryland. The officer on duty told Allen that President Reagan was just boarding the White House helicopter to return to Washington.

"Better get him off," Allen instructed.

A minute later Reagan was on the telephone. Allen could hear the *whump-whump* of helicopter blades rotating in the background.

"Yes?" Reagan said.

"Mr. President, the Israelis just took out a nuclear reactor in Iraq with F-16s," Allen said.

"What do you know about it?"

"Nothing, sir. I'm waiting for a report."

"Why do you suppose they did it?" Reagan asked, then, not waiting for a response, answered the question himself. "Well," he shrugged, "boys will be boys."

If nonplussed, Reagan's response was not surprising. The president had been fiercely pro-Israel and a staunch foe of anti-Semitism his entire professional life, dating back to his days as a Hollywood liberal. An FBI dossier on Ronald Reagan from the post–World War II years recounted an episode at a Hollywood party when Reagan nearly came to blows with a guest who had accused the Jews of war profiteering. His support for Israel was one of the bedrock convictions that survived Reagan's political transformation from liberal Democrat to conservative Republican during the 1950s. It was made all the stronger by what even his friends confessed was a piecemeal understanding of the complex history of the Middle East. He only knew that Israel's interests were the United States' interests. The president's well-known loyalty to Israel did not extend to every member of his administration, however.

The Israelis' audacious military attack sent shock waves from the White House to Foggy Bottom and across the Potomac to the

Pentagon. Secretary of Defense Caspar Weinberger was given a rundown of the raid at Defense's Monday morning meeting held at the Pentagon. Weinberger had mixed feelings about the attack. On the one hand he was generally sympathetic to Israel and her security needs. Nonetheless, he questioned the wisdom of this unilateral action without prior notification to the United States and in direct violation of the U.S. Arms Export Control Act, which stipulated that all U.S.-supplied weaponry be used for defensive purposes only. Even more troubling was how the other Arab nations would respond. The United States was selling a tremendous amount of military hardware to nearly all the countries in the region, including Jordan, Iraq, Egypt, and Saudi Arabia. In fact, Saudi Arabia would quickly argue that Israel's violation of its airspace during the mission justified extending the sale to the sheikh of two more AWACS. Every one of these countries was restricted by the same conditions of sale. Weinberger worried that if the United States failed to sanction Israel, would that send a message to the Arab states that they, too, could now use their American-supplied arms to attack Israel—or one another?

Later that morning, at the regular Monday meeting of Reagan's NSA team, the White House chieftains were all in high dudgeon over what they considered unwarranted Israeli aggression. Secretary of State Alexander Haig, normally a staunch Israeli supporter, called the raid "reckless." He argued that the United States had no choice but to formally protest the attack and should impose sanctions on Israel immediately. Reagan's chief of staff James A. Baker agreed that "some kind" of sanctions were called for. Israel had violated the strict conditions of the sale of military hardware. Its raid was clearly an offensive action.

Finally, it was Weinberger's turn to speak. He told Reagan that, in his estimation, they had no other choice but to suspend the remainder of the F-16 sale to Israel—at least, temporarily. Four new planes were sitting in the hangar at General Dynamics awaiting delivery even as they spoke.

Throughout the meeting, Reagan listened patiently and said nothing. Privately, however, the president could not see what the big deal was. In fact, at one point he looked across the table at Richard Allen and rolled his eyes, as if to say, "Oh, brother!"

But in the end, Reagan agreed to go along with the recommendations of his advisers. The State Department would immediately release a statement strongly condemning Israel's "aggressive" and "unprovoked" attack on Iraq. Invoking the U.S. Arms Export Control Act, Secretary of State Haig announced that the United States was immediately suspending any further sales of F-16s to Israel, including the four planes sitting at General Dynamics.

Haig and the administration bureaucrats had had their way: they had slapped Israel on the wrist—hard. But prevailing opinion concerning Israel's action was hardly lockstep inside Washington's corridors of power. As early as the first Monday morning meeting of Defense, it was clear that many around the table, including ranking members from the Joint Chiefs, had at least a grudging admiration for the boldness and the remarkable precision of the attack. Others, like Richard Perle, at the time an assistant secretary of defense under Weinberger, disagreed with the decision to censure Israel and were outspoken in their support of the Israeli action. Perle thought it a great act of antiproliferation, the exact thing the United States should be doing more of.

Indeed, late Monday afternoon, when Richard Allen brought in the highly classified KH-11 satellite photos of al-Tuwaitha to show to the president, Cap Weinberger, and the Pentagon generals, the reaction of the group was amazed silence. Clearly seen were the surrounding fences, the outlying buildings, the main gate, the guard towers—everything perfectly, immaculately intact. But in the middle of it all stood a deep, gaping hole: the site of the former Osirak reactor, utterly and surgically obliterated.

Reagan studied the photographs, then finally said what many present had silently been thinking.

"Okay, yeah, yeah, I see," the president said, referring to the putatively damning evidence of Israel's perfidy. "But what a terrific piece of bombing!"

Monday morning, July 8, Hagai Katz headed with his family to the countryside to celebrate Shavuot with tractor rides, picnicking on hay bales, and listening to the outdoor songfests. Radios everywhere played traditional Israeli folk songs and music. Katz and his wife were sitting on a hay bale, nibbling on their packed lunch, when a man picnicking next to them leaned toward Hagai and announced, "Did you hear? We just blew up Iraq's nuclear reactor!"

Katz couldn't believe his ears. What the hell? he thought. They had all just been resworn to secrecy the night before. How did the news get out?

Zeev Raz and his wife had elected to celebrate Shavuot by staying home and relaxing for the first time in months. At 3:30 in the afternoon, the music programming on the radio was interrupted by a special announcement: the Israeli Air Force had successfully destroyed Iraq's Osirak nuclear reactor, a critical part of the country's plans to produce an atomic bomb.

Raz was shocked by the announcement. But he said nothing.

The last eighteen months had been a hard time for Raz and his wife. He had been gone long hours, involved in something intense and serious, something she had been completely cut out of. She worried, and she hated the many, many hours he spent away from home. Something had been eating at him. She could see it in his face, feel it in his body. Especially in the last few weeks. But he wouldn't tell her. She felt cheated and, gradually, resentful.

Now, finally, it was clear. With the announcement, she knew how important her husband's work had been. Her eyes filling with tears, she walked to Raz and hugged him.

"Now I know why you've been working so hard all this time," she whispered.

Raz was relieved that his wife finally knew, but like Katz, he wondered how word had gotten out. Who had leaked?

Two hours after Raful Eitan called to tell him the attack had been a success, Begin had phoned the United States ambassador to Israel, Samuel Lewis, and informed him that the Israeli Air Force had just bombed Iraq's Tammuz nuclear reactor.

"You don't say?" Lewis deadpanned. He then immediately telephoned the State Department in Washington, D.C., and relayed the news.

Begin was bound by honor and diplomacy to inform Israel's closest ally of the attack. But Begin secretly hoped that the United States would break the announcement to the world, thereby implicitly involving itself in the attack and taking some of the heat off Israel. Moreover, the Israeli parliament, when informed of the raid, made Begin promise that no one in the government would talk about the mission unless word of it broke first in an outside source. But the prime minister was impatient to get the story out—first, because the raid had been an unqualified success; and second, because he wanted to take the offensive position to blunt the international outcry he knew would be forthcoming. In fact, he had already ordered his press secretary, Uri Porath, to draft a press release detailing the mission. All Monday morning the prime minister waited for news to break in the world press. But the Reagan administration had refused to bite. They were staying way away from this one.

Finally, around noon, Begin saw an opening. During a public debate in the Jordanian Parliament broadcast over the nation's airwaves, the Jordanian prime minister accused Israeli planes of taking part in the Iran-Iraq war. The prime minister was not alluding to the attack on Osirak, since even Iraq was not sure at the time of the nationality of the fighter planes that had bombed al-Tuwaitha, but Begin heard what he wanted to hear: Israel was being named as the perpetrator of the air strike.

"Release the statement," he ordered his press secretary.

Hours later there had still been no announcement. Begin, irate, called in Porath.

"Didn't you release that statement?" he snapped.

"Yes, I called it in hours ago."

Begin stormed to his telephone.

Just before three o'clock, Emmanuel Halprin received a phone call at his staff office at KOL YISRAEL, the Voice of Israel. The call was from his uncle, Menachem Begin.

"Yes, Uncle?" Halprin answered dubiously, wondering what was going on.

"Did you receive a press statement from my office this afternoon?" Begin snapped.

"Well . . ."

In fact, Halprin had been puzzling over a bizarre "press release" talking about a bombing of Iraq's nuclear reactor, which had been dictated over the phone to the radio station earlier, supposedly from the prime minister's office. Because of the Shavuot holiday, the radio station had only a skeleton crew. The announcement sounded too incredible. Maybe the receptionist had been taken in by a prankster. Halprin had decided to hold it.

"We thought it was a hoax," Emmanuel said. "Is it real?"

"Yes, dammit," Begin snorted. "Get it broadcast. Now!"

Emmanuel hung up the phone and immediately walked the

statement into the announcer's booth. Programming was interrupted for a special announcement: "The Israel Air Force yesterday attacked and destroyed completely the Osirak nuclear reactor which is near Baghdad. . . ."

That Sunday, the cavernous lobby of Baghdad's Palestine Hotel was crowded, as it was on many such weekends that summer. The sea of smiling faces and tailored dark suits perched on gilded chairs or standing on the plush red carpet telegraphed the universal sign of successful salesmen everywhere. In this case, they belonged to international arms dealers, gathered to ply their latest high-tech weapons systems, bombs, torpedoes, radars, tanks, and mines to the world's biggest buyer. They passed out brochures in French, Russian, English, and Serbian, touting comic book–sounding names like Chinooks and Big Mothers and Phantoms. For these men it was business as usual on Sunday, though their hosts, more than one salesman remarked, seemed unusually distracted this evening. The normally obliging Iraqi ministers had either canceled meetings or left appointments early. The Iraqi media gave not a hint of anything untoward.

By late Monday afternoon, however, the dealers had all learned the truth of their Iraqi hosts' sudden anxiety. The rumor of the destruction of Osirak ran throughout the hotel. The salesmen were quick to commiserate. It was a dangerous and unjust world they all lived in, they consoled. But then, all was not lost. Iraq could always rebuild. After all, the oil was still flowing. Now more than ever, Iraq needed the latest in Western technology and defenses. In fact, as the French arms dealer pointed out, they were selling an entirely new generation of advanced early-warning radar defenses. France could deliver within the month.

BLOWBACK

No good deed will go unpunished.

—ANONYMOUS

Monday morning, June 8, 1981, Khidhir Hamza drove his Passat to al-Tuwaitha, hoping to discover what was going on. But as the director pulled up to the main gate, he was stopped by grim-faced Mukhabarat guards armed with AK-47s. They checked Hamza's identification, then informed him that no one was being allowed inside the compound. An Iraq Air Force explosives team was still securing the grounds.

It would be several days before Hamza and his colleagues were allowed back into the facility. As he walked the familiar pathway to his office in the AE administration building, the scientist saw scores of bomb specialists, sappers, uniformed security, construction laborers, and dark men in blue suits and fedoras combing the area. The uniformed men looked gloomy and nervous, demoralized. The Iraqi army had failed to bring down even one fighter plane. There was no evidence that an enemy plane had even been hit. And not one MiG had been scrambled. It was a repeat of the same sorry performance against the Iranians nine months earlier.

Heads would roll, they knew they could count on that. Indeed, when Saddam Hussein learned that the antiaircraft units were at dinner at the beginning of the raid, he had the commander of the AAA batteries taken out and shot.

Hamza walked straight to the crater that had been Osirak. He circled the reactor. The spectacular dome was completely gone. The pool below, where the reactor fuel rods were cooled, was filled with twisted steel and broken concrete. The enriched uranium already exported by France and stored underground next to the neutron guide hall was unharmed. The air force investigators found an unexploded two-thousand-pound bomb in the hall's concrete-encased tunnel. At first the sappers thought that it was a booby trap, a bomb equipped with a delayed fuse to blow up innocent civilians. The explosive, of course, was one of Spector's misses. It had been dropped at too low an altitude, and so the fuse had not had time to arm itself.

Typical of Iraqi culture, especially a totalitarian state in which information was hoarded and manipulated to create fear in the general population, fantastical rumors raced through the NRC's workers, even the educated scientists. The day before the raid, suspicious-looking men had supposedly been spotted lurking about the neutron guide hall in a van. They had been delivering radiation detection equipment, but later, it was said, an electronic guidance transmitter had been found inside the hall. The French and Italian workers had all suddenly been called back to the foreign housing compound just hours before the attack. One Frenchman had refused to leave—Damen Chaussepied, the technician who had been killed in the explosions. Iraqi cooks in the foreigners' compound reported overhearing loud arguments between the workers that night. Of course, nothing came of the rumors. But they underscored the uncertainty of the center's employees. Would the French return? Would they rebuild? Did they still have jobs? Would the Israelis return and bomb the rest of the plant?

One fact was incontrovertible: Osirak was no more. As the French technician Jacques Rimbaud told the Paris press the day after the raid: "The central building is destroyed; the anti-atomic shelter has vanished. If they want to resume work, they will have to flatten everything and start from scratch."

The storm of indignation Menachem Begin had been anticipating ever since his phone call to the American ambassador Sunday night hit like a blizzard Tuesday, June 8. The U.S. State Department's censorious release chastising Israel on Monday was but a snow flurry ahead of the main front. France, not surprisingly, was outraged by the destruction of its nuclear reactor and the end to so many lucrative contracts. And once again, the country was embarrassed by the renewed worldwide focus on its involvement in Iraq's nuclear aspirations. French president François Mitterrand, Peres's good friend, rebuked Israel. "Any violation of the law will lead to our condemnation," he announced to the French populace. "Whatever may be our feelings for Israel, this is the case now concerning the intervention decided by Israeli leaders against Iraq, which has led to the death of one of our compatriots." This last reference was to the nation's new hero, Damen Chaussepied, the technician killed during the bombing. Immediately, the foreign office ordered home 115 nuclear scientists and engineers from al-Tuwaitha, leaving 15 behind to help ascertain whether there was danger from radiation leaks.

The foreign minister Claude Cheysson charged that the attack was "unacceptable, dangerous and a serious violation of international law." "I am saddened," he told reporters on June 9. "This government has a great deal of sympathy for Israel, but we don't think such action serves the cause of peace in the area."

But France was not content with verbal condemnation. Feeling betrayed by Israel, high-placed French officials and members of the country's intelligence service began leaking classified information to the world press about the secret reactor and plutonium reprocessing facilities the country had helped Israel construct in Dimona decades earlier. The Arab states in the region began clamoring for a full investigation of Israel's nuclear capabilities and her immediate disarmament.

Even Britain denounced the bombing. Usually conservative Prime Minister Margaret Thatcher declared, "Armed attack in such circumstances cannot be justified. It represents a grave breach of international law." Meanwhile, British intelligence officials, still stinging over being shut out of full access to KH-11 photographs while Israeli agents blithely rifled through whatever film they wanted, immediately suspected Israel of using smuggled high-resolution satellite surveillance shots to help target Osirak. The British promptly lodged their complaints with CIA, in effect, telling Casey: "We told you so." The complaint prompted the CIA director to initiate the confidential, high-level investigation into Israel's ability to access restricted satellite imagery, a study that revealed a complete breakdown in the monitoring system, allowing Israel virtually full run of the satellite-imaging henhouse, as it were. According to writer Seymour Hersh, one angry Pentagon official declared at the time, "The Israelis did everything except task the bird," referring to the ultimate ability to select targets and, thus, reroute the orbiting patterns of the satellite in space. In the end, Casey continued to allow Israel access to KH-11, but with the original 1979 restrictions of the Carter administration firmly back in place.

Most of the First World nations around the globe also joined France and Britain in lambasting Israel. Japan stated, "Israel's action cannot be justified under any circumstances." The West German foreign ministry said it was "dismayed and concerned" by the raid. The Greeks called it "unacceptable." Even the Argentine

Foreign Ministry declared Israel's action "a threat to the peace and security in the Middle East."

In a stinging blow to Prime Minister Begin, and to Ivry, who had been surprised by the bitterness of the State Department's response, the United States populace seemed to be siding against Israel as well. A *New York Times* editorial on Tuesday following the raid excoriated Israel, charging, "Israel's sneak attack on a French-built nuclear reactor near Baghdad was an act of inexcusable and short-sighted aggression." *Time* magazine maintained that the attack endangered the historic gains of the Camp David Accords, insisting that "Israel has vastly compounded the difficulties of procuring a peaceful settlement of the confrontation in the Middle East."

And in a historic turnabout, the U.S. ambassador to the United Nations, Jeanne Kirkpatrick, blistered the Israelis, calling the raid "shocking" and going so far as to compare it to the recent Soviet invasion of Afghanistan. The United States also approved the passing of United Nations Resolution 487, which strongly condemned "military attacks by Israel in clear violation of the United Nations Charter and the norms of international law" and called for Israel to make "appropriate redress" to Iraq.

The entire Arab bloc damned the attack on Iraq's sovereign territory. Egyptian president Anwar Sadat, who had just met publicly with Begin in the Sinai that weekend to further the road to peace, was furious. Privately, he told colleagues that he felt ambushed by the raid, made to look as though he were somehow complicit in the attack. The Egyptian parliament requested the United States to reassess its military aid to Israel.

Back in Baghdad, Hussein and the Iraqis took full advantage of their new role as aggrieved victim. Ironically, Iraq was not even sure who had bombed Osirak until Begin released the news bulletin on Monday. Hussein found himself deluged with messages of outraged support from Kuwait, Jordan's King Hussein, the PLO, Syria, which decried the "Zionist enemy aggressions," the United Arab Emirates, Bahrain, Morocco, and even Kenya, which called the raid "indefensible." Libya's loose cannon, Col. Muammar Qaddafi, called on his Arab brothers to blow up the Israeli nuclear reactor in Dimona in revenge.

On June 23, Saddam Hussein finally addressed the public for the first time since the raid. He called on "all peace-loving nations of the world to help the Arabs in one way or another acquire atomic weapons" in order to offset Israel's "nuclear capability." Hussein, however, quickly distanced himself and his Ba'th Party from the disaster at al-Tuwaitha, accusing the French of complicity with Israel and denouncing his own Atomic Energy administration for lax security and failing to anticipate a hostile military attack—even though Hussein himself had vetoed plans early on to construct the reactor belowground.

Begin closely monitored the foreign reports and media stories. He was furious over the international outcry. Nearly beside himself with indignation, against the advice of his advisers, the prime minister immediately took to the offensive. On Tuesday, flanked by Eitan, Ivry, and Saguy, Begin, looking every bit the unbowed fighter, held a fiery press conference to rebut the global censure, defiantly declaring that Saddam Hussein had already butchered his closest colleagues and would have had "no hesitation in dropping three or four or five of those bombs on Israel." Not for a second, Begin insisted, did he regret his decision.

"Israel has nothing to apologize for," he snapped into the microphone in front of him. On the contrary, he exclaimed, "I feel like a man who's left prison. I feel like a free man!"

Indeed, the raid solidified Begin's popularity in Israel as a defender of the nation. Before the June 7 raid, a poll in the *Jerusalem Post* showed Shimon Peres's Labor Party continuing to hold a steady lead over Begin and the Likud Party. By the evening of election day, June 30, 1981, Begin and Likud had been swept to victory, ushering in a political and cultural revolution. For the first time in the history of Israel, the Socialist-Zionist alliance of European and American Jews that had guided the nation from its inception had been repudiated, replaced by the most hawkish government ever assembled in Jerusalem. Indicative of its conservative bent was Begin's appointment of Ariel Sharon as defense minister and Yitzhak Shamir as foreign minister.

Even the one true political setback proceeding from the raid, the United States' suspending the sale of the F-16s, seemed to turn around to Begin's side. At a June 16 news conference following Osirak, President Reagan seemed anything but angry about the preemptive strike. When reporters asked about his reaction to Israel's refusal to sign the Nuclear Nonproliferation Treaty, Reagan responded, nonplussed, "Well, I haven't given very much thought to that particular question there," then added that it was difficult "to envision Israel as being a threat to its neighbors." By September 1, 1981, the sale of F-16s to Israel was quietly resumed.

Perhaps in somewhat the same way that the United States' stunning victories in Afghanistan more than twenty years later would embolden George W. Bush's administration to launch an offensive against Iraq, so Osirak in 1981 precipitated a series of bold political and military moves by a newly confident and invigorated Menachem Begin. In the prime minister's estimation, Osirak had given him a mandate to quash all of Israel's enemies once and for all. Heartened by the success of the attack, Begin and Sharon were determined to drive Yasser Arafat and the Palestine Liberation Organization out of Beirut. Within the year, Begin made his fateful decision to invade Lebanon and lay siege to Beirut. In the bloody chaos that followed, Sharon would fatally miscalculate the military's control over Israel's Christian Phalangist allies. While under the supposed protection of the IDF, a renegade Phalangist battalion swooped into the densely populated Palestinian refugee camps at Lebanon's Sabra and Shatila farms and massacred 750 men, women, and children while Israeli military forces stood by impotently and watched. The resulting outcry both worldwide and within Israel nearly toppled Begin's government and ultimately led to Sharon's dismissal and years of bitter political exile.

The June 7 raid unleashed other unexpected aftershocks for Begin as well. During his monumental press conference two days after the attack, the prime minister, in his zeal to praise the IDF, bragged that the Israeli F-16s had destroyed a secret facility buried "forty meters beneath the reactor"—a hidden plant for the production of atomic bombs. There had been no such secret facility beneath Osirak—the closest thing to it might have been the nuclear guide hall, the experimental laboratory to investigate the property of neutrons that extended beneath the ground from the reactor. To the horror of Mossad's Hofi and the IDF's Saguy, what Begin was describing did in fact exist, but not at al-Tuwaitha. He had confused Osirak with Israel's own supersecret A-bomb plant 120 feet belowground at Dimona.

The prime minister's press office tried to recoup the next day by explaining that Begin meant to say "four" meters under the ground, not "forty," when describing the facility at Osirak. But the damage was done: the CIA, which had been deeply suspicious of the Israeli nuclear facility for years, was more than intrigued by the "misquote" and knew immediately what Begin had done. He had blown his government's secret operation. Mossad director Hofi was furious. He had spent a year in the doghouse being punished by Begin over his opposition to Osirak. Now Hofi had no sympathy for Begin. Two weeks after the PM's press conference, the Mossad director granted a rare interview to the Israeli press. In it, Hofi complained bitterly about "politicians" who were compromising the nation's secret intelligence and undermining the security of the state. There was no doubt within the Israeli political elite about whom Hofi was referring to.

The Arab nations capitalized on the opportunity to refocus world attention on Israel's atomic aspirations. Joined by France, Germany, Italy, the United States, and the United Nations' IAEA, they demanded an investigation into Israel's nuclear capabilities and a full disclosure by the nation concerning its production and distribution of any and all atomic weapons. Begin adamantly refused. The result was an avalanche of negative press that portrayed Israel as an aggressor, a dangerous nuclear power, even a pariah state in the Middle East. The flap set back Israel's foreign relations with many European countries for much of the '80s. The country's opposition Labor Party, headed by Peres, accused Begin of setting the nation on a suicidal path of global isolation.

As for Hofi, his opposition to the raid cost him his job. His deputy director, Nahum Admoni, had disagreed with his boss about Osirak from the beginning, arguing at one point that even if destroying the reactor was not a matter of life and death, it would teach "any other Arabs with big ideas a lesson." The onetime close friends gradually became hardened political enemies. A majority of the Mossad sided with Admoni, as Hofi became more and more isolated at the top. A little more than six months after the raid, Yitzhak Hofi, who had headed the secret intelligence service since 1975, was squeezed out in January 1982, replaced by Admoni.

Inside Iraq, the June 7 attack brought to a screeching halt the nation's secret plans to use plutonium extraction from spent reactor fuel rods as a route to creating an atomic bomb. But it did not end

Saddam Hussein's dreams of nuclear dominance. There were, after all, other ways to obtain enriched uranium.

By November 1981, the Nuclear Research Center was already beginning to retool to accommodate new technology. The Osirak reactor was cleared of rubble, but remained for the most part a neglected crater in the middle of the compound. Hearing about deteriorating morale among his scientists after the raid, Hussein made a second visit to Atomic Energy, arriving early in the morning, dressed for battle—black beret, olive green army togs, and a holstered pistol on his hip.

"If you are scared now, how do you think you would do in a real shooting war?" Hussein excoriated Khidhir Hamza and the hundreds of scientists and employees who had been rounded up to be lectured to by the Great Uncle.

"You think the Iranian mullahs are weak? You think those bearded fanatics will give up?" he snapped. "No, they are not weak. They will never give up. They can't wait to die for Allah!"

Did they want to be oppressed by religious fanatics who would force their women to cover themselves in veils? No, they had to fight.

Before departing, Hussein offered his carrot, announcing he was leaving behind twenty-six brand-new automobiles as a gift. They could decide among themselves who best deserved them. And then he was gone.

Soon after, Jaffar Jaffar, the former head of AE, was suddenly released from prison and transferred back to al-Tuwaitha. Under the umbrella of Atomic Energy, a new, independent, top-secret department was formed called the Office of Research and Development. Headed by Jaffar Jaffar and staffed by al-Ghafour and Khidhir Hamza, reporting directly to Hussein's brother-in-law Barzan al-Tikriti, the organization would follow simultaneously two new routes to producing enriched uranium: the first technique was use of centrifugal technology; the second was use of magnetics.

The centrifugal process was used by Pakistan in that nation's atomic weapons program. Complex and expensive, the technique converts uranium ore into a uranium gas, then separates out small U-235 atoms from U-238 by repeatedly spinning the uranium-compound gas inside a rapidly rotating cylinder, enriching lighter uranium at the center. The second process, pioneered by the Manhattan Project scientists working on the Little Boy bomb during World War II, creates enriched uranium by using powerful electromagnets. Both processes were incredibly technical, time-consuming, and required a great deal of highly sophisticated and expensive scientific equipment, much of which was on the IAEA's list of proscribed technology.

Already facing such formidable obstacles, the nuclear scientists' work was hampered even more critically by Hussein's bloody, costly war with Iran, which continued to drag on for most of the eighties. Both sides had become mired in demoralizing, WWI-style trench warfare along Iraq's eastern borders, neither side giving nor gaining ground. In the end, the conflict would cost both nations over a million lives, seriously damaging Iraq's once-thriving economy. The conflict would also drain off many of the Nuclear Center's resources and funding. Procuring the necessary technologies to implement either centrifugal or magnetic processes for enriching uranium tooks years sometimes. Meanwhile, well before Iraq was ready to produce its first weapons-grade uranium and begin creating its own atomic bomb, Saddam Hussein invaded the tiny principality of Kuwait on the eastern border of Iraq. Hussein had long claimed Kuwait and its wealthy oil fields—arbitrarily sectioned off of the country by the British in 1932—as a part of the Iraqi empire. Convinced that the United States would go along with the annexation, Hussein overran the undefended state in days, his troops looting and terrorizing the mostly upper-class citizens of the kingdom all the way to Kuwait City. President George H. W. Bush responded by piecing together the now-famous worldwide coalition that swept Hussein's armies out of Kuwait and the southern Shi'ite territories of Iraq within weeks. During the nearly monthlong bombing campaign that preceded the ground offensive, Khidhir Hamza's Nuclear Research Center at al-Tuwaitha was heavily targeted by Allied sorties.

By the end of the war, the once-proud campus was a moonscape of rubble and destruction. What the Israelis had done to the Osirak reactor, coalition bombers had done on a mammoth scale to the entire complex. The buildings and laboratories were either leveled or gutted, the towers toppled. In the peace that followed the Gulf War, United Nations weapons inspectors would harry and hound the remnants of Saddam Hussein's secret nuclear weapons programs underground until 1998, when the inspectors were pulled out prior to Coalition bombing ordered by President Clinton in the wake of Hussein's refusal to cooperate with U.N. strictures. The atomic programs would continue, but at a pace nowhere near what Hamza and his colleagues had overseen during the heady days at al-Tuwaitha. The seeds of destruction of Iraq's nuclear ambitions had been sown on June 7, 1981, and Hussein was never to see his dream become a reality. In the passing years, many of Iraq's scientists would defect, seeking asylum in the West. Khidhir Hamza fled Iraq in 1995, finally moving to the United States and being granted asylum after months and months of hiding from Iraqi agents and negotiating with the CIA. Eventually, Hamza would renew his teaching career and become one of the outspoken Iraqi exiles, lobbying against Hussein's regime and warning of the nation's ongoing nuclear, biological, and chemical weapons programs.

On June 7, 2001, the eight mission pilots along with the two backup pilots and all their wives gathered together for the first time in two decades to mark the twentieth anniversary of the raid on Osirak. They met for dinner at the beautiful, airy home of Iftach Spector in a bucolic neighborhood some twenty miles outside Tel Aviv. Many of the pilots had not seen one another in years. Generals Rani Falk and Amos Yadlin were still in active service in the IAF. Ilan Ramon was on active duty, but now reassigned as Israel's first astronaut, training with the Columbia Shuttle crew in Houston, Texas. Others, like Relik Shafir, remained in the reserves. Raz, Hagai, and Nachumi had retired and were working in the public sector.

The years after the raid had brought them all an odd kind of anonymous celebrity. Their deed was known throughout the world and lionized in the Israeli press, but because of the military's insistence on continued secrecy about the mission, which, at least officially, remained classified, few outside the IAF, wives, and family knew who they were. But the pilots were satisfied with the respect and appreciation of those colleagues and high-level professionals who understood exactly what they had accomplished. For months, in fact, the Pentagon had refused to believe that the pilots had truly flown to al-Tuwaitha and back without refueling. There were rumors that the F-16s had actually been refueled over Saudi Arabia by tankers disguised as commercial planes. Other fantastical stories had it that the pilots had flown the entire distance grouped so tightly together that ground radar would mistakenly "read" them as a commercial flight. Amazed by the accuracy of the targeting, the Arab nations charged that Mossad spies had planted secret homing devices in the reactor. The pilots would quietly shake their heads and laugh.

Katz, in fact, was given the task of walking a skeptical U.S. Air Force captain sent from the Defense Department and an official from General Dynamics step-by-step through the entire mission in order to demonstrate how they had pulled off the attack on "one tank of gas."

Now at Spector's home, the pilots reminisced and swapped stories about the raid and the training—events they had forgotten, things they had never told. Nachumi recalled that the day after the raid, when the ground crews had rolled out the F-16s for a routine maintenance check, not one of the eight planes had started. All had mechanical failures.

"Who says planes do not have souls," Nachumi declared.

For the first time, too, the wives shared their experiences: who had known about the raid beforehand, and who had not. After catching up on old business and family history, the men gathered for a video presentation assembled by Amos Yadlin and Relik Shafir. For the first time in twenty years, they watched videotaped highlights of the attack captured by their nose cameras. Many of the men had forgotten over time the terrible sounds of the AAA fire, the streaking SA-7s, and the hysterical radio chatter of the ground crews below.

During the evening, Katz found himself thinking more than once about Spector. He had felt that, knowing the commander the way he did, the bitter memory of missing his target that day would have eaten at him for the last twenty years. Those misses were indeed still on Spector's mind. He presented the squadron with a bottle of champagne and a card he had made up in his immaculate block print. It read: "Our mathematical magic: 7 divided by 8 = 100 percent." Even though only seven pilots out of eight hit their target, the squadron together was perfect. It was Iftach Spector's way of honoring his fellow fliers and acknowledging his personal failure. Then, that night, for the first time in twenty years, Spector explained to the men what had happened, that he had been sick and dizzy and perhaps had blacked out. He could not be sure. Two decades had passed and he still could not remember.

And he could not forget.

Rani Falk asked why he had waited twenty years to tell them.

Spector just shrugged. The revelation only made more acute a sense of guilt Nachumi had carried for two decades. At the most critical moment, he had put even more pressure on his unhealthy wingman by so perilously closing up the attack formation and then not focusing on his position at all—at the very moment he may have caught blackout.

General David Ivry did not make the pilots' reunion. In 1998 the former head of IAF had been appointed Israel's ambassador to the United States, and that summer of 2001 he would be wrapping up his tenure in Washington, D.C., where he had become one of Israel's most popular and respected officials. Indeed, a Who's Who of Washington turned out at his retirement party the following April, held inside the tented courtyard of the great gray-stoned fortress that is the Israeli embassy. Tucked away on leafy Consulate Row in D.C.'s Van Ness district, surrounded by a tall iron fence and patrolled by many guards, some in uniform, some not, the three-story estate sits on a corner lot just down the street from the shuttered Ethiopian embassy. Alighting from their limousines that night were Vice President Richard Cheney, Deputy Defense Secretary Paul Wolfowitz, Senator Lindsey Graham, publisher Mort Zuckerman, and, seemingly, an entire delegation from the neoconservative think tank American Enterprise Institute, whose membership, like its famed conceptualists Wolfowitz and then Defense Advisory Board chairman Richard Perle, had been inspired by Ivry's groundbreaking preemptive strike on Osirak, holding it up as the prototypical example of the kind of first-strike "defensive" action they argued should form the basis of a new, aggressive American foreign relations philosophy, especially regarding Iraq.

Like the ambassadors before him, Ivry had taken up his post in the large corner office of the embassy. Here he had spent much of the last week packing up his personal belongings, including his beloved collection of models of Israeli fighter planes, and various mementos and honors of his ambassadorship. Cherished among these was the framed, enlarged print of a postcard he had hung on his office wall, across from his mahogany desk, where he could see it whenever he looked up. For the general, who had been surprised and stung by the criticisms of the Osirak mission hurled at the time by Secretaries Haig and Weinberger and Ambassador Kirkpatrick, the missive had a special meaning. The homemade postcard was actually a blowup of a satellite photograph (ironically, the KH-11 spy satellite) of al-Tuwaitha taken from space in the days after the Israeli attack in 1981. Outlined by a huge rectangle of twenty-foot-high concrete walls and barbed wire was, unmistakably, the bombed-out crater of an immense dome, its once-shiny aluminum cupola crumpled, the concrete crumbling. It was all that remained of Osirak after the Israelis were through with it.

Handwritten at the bottom of the photograph was a short note, penned a week after the coalition's successful invasion of Iraq in 1991. It read:

"With thanks and appreciation. You made our job easier in Desert Storm."

It was signed: "Dick Cheney."

ACKNOWLEDGMENTS

First things first: I need to thank Don and Kathy Hutchison, without whose belief, advice, and support—literally—from the earliest days, this book would not exist. As for the writing itself, I owe much to the perspicaciousness and craftsmanship of my agent, David Halpern, of The Robbins Office, and of my editor at Broadway Books, Charles Conrad. David not only did all the wonderful things agents do—like getting me paid—but also functioned as a first editor, advising me on style, pace, and organization. Charlie proved to be a superior editor, not only in helping me frame my tale but also by suggesting wonderfully imaginative ways to make a sometimes complex military story eminently readable. The "countdown" sequences that precede each chapter were Charlie's genius.

I would especially like to thank Tal Yanai for his always-dependable research, Hebrew translations, and deep knowledge of Israeli history and politics. Without Tal and his wife, Dr. Jamie Miles, I would have had a difficult task absorbing the complexities of Israeli culture and the national character. Likewise, thanks to Hossan Hamalawy, an Egyptian stringer for the *Los Angeles Times*; Hossan found and interviewed an Egyptian student who'd befriended and admired Hussein while he was in exile in Cairo.

Like most creative enterprises, a book is also a team effort, and I would never have succeeded without the invaluable aid of Mark Regev, the press counselor at the Israeli embassy in Washington, D.C., who smoothed the bureaucratic waters on both sides of the Atlantic; editorial assistant Alison Presley, who made everything so much easier; Bob Gaston, for astute military and legal insights; Geoff Miller, for consistently good counsel and taste; famed *Time* photographer Ben Martin, who shot my jacket portrait; Major Janske of the California Air National Guard, 144th Fighter Wing, for an exhilarating up-close look at the F-16 Fighting Falcon; my wife, Ann, and the girls, Wren and Kelsey, for allowing me to commandeer the dining room table for twelve months; and, finally, the guys at The Farm.

NOTES

Author's note: The first time a work or interview is referenced, I have included all relevant information. All subsequent references to the work are sourced by the last name of the interviewee or author only.

PROLOGUE: THE ROAD TO BABYLON

1 **"General David Ivry's wife . . . 'Shalom.' "** The episode was told to me by former ambassador David Ivry during my first interview with him in September 2001. Other observations come from discussions with Mark Regev, the Israeli embassy's communications director.

3 **"A staff car . . . planning this mission."** Ivry.

3 **"Waiting for him . . . minus six hours and counting."** Ivry; also *The Life and Times of Menachem Begin,* Amos Perlmutter (Doubleday: New York, 1987).

4 **"The planes below stood menacingly anonymous . . . either side of the fuselage."** There is very little written about the Osirak raid. An early and almost completely overlooked book about the mission is *Bullseye One Reactor* by Dan McKinnon (House of Hits: San Diego, 1987), a former USAF pilot who wrote a remarkably well-researched telling of the attack, especially considering the muzzle the IDF had kept on the media and a lack of access to the pilots, who are referred to by pseudonyms. The out-of-print book was recommended to me by Amir Nachumi, who shared many of his early recollections with McKinnon.

4 **"As Ivry walked up . . . Israel for trial."** Perlmutter.

5 **"Raful's son, Yoram . . . seven days of seclusion."** McKinnon; Ivry; interview with mission pilot Doobi Yaffe.

5 **"Eitan caught Ivry's look . . . Eitan said."** Conversation is a reconstruction through discussions with Ivry and Regev.

5 **"Eight pilots would have to fly . . . nuclear reactor."** Interviews with mission pilots Zeev Raz and Hagai Katz.

7 **"The modeling experts . . . antiaircraft fire."** Interview with mission pilot Gen. Amos Yadlin.

7 **"The two generals moved . . . 'the mission blown.'"** Interview with mission pilot Amir Nachumi and Ramat David commander Iftach Spector; McKinnon.

8 **"The crew chiefs . . . four planes to a line."** Ivry, Yaffe, Nachumi.

TERROR OF THE TIGRIS

11 **"Before the birth . . . since the earliest days of the regime."** The section of the chapter dealing with Saddam Hussein's early life and rise to power was compiled from interviews, background research, and the invaluable help of several good books on Iraq and its Ba'thist president: *Instant Empire: Saddam Hussein's Ambition for Iraq,* Simon Henderson (Mercury House: San Francisco, 1991); *Saddam Hussein: A Political Biography,* Efraim Karsh and Inari Rautsi (MacMillan, Inc.: New York, 1991); *Saddam Hussein: The Politics of Revenge,* Said K. Aburish (Bloomsbury: London, 1999); *The Continuing Storm: Iraq, Poisonous Weapons and Deterence,* Avigdor Hasekorn (Yale University Press: New Haven and London, 1999); J. Snyder, "The Road to Osirak: Baghdad's Quest for the Bomb," *Middle East Journal,* vol. 37, no. 4, 1983; interview with Sayyed Nassar, an Arab nationalist who befriended Saddam Hussein when the Ba'thist was in exile in Cairo in the 1960s.

19 **"Halfway across the world . . . it already had close to one hundred of them."** This section of the chapter, recounting Dr. Khidhir Hamza's experiences in Iraq's Nuclear Research Center, is based on both a 2002 interview with Dr. Hamza outside Fredericksburg, Virginia (hereafter referred to as "Hamza"), and his autobiographical book about his adventures, *Saddam's Bombmaker* (Scribner: New York, 2000).

23 **"The year was 1956 . . . except Saddam Hussein."** The section recounting Israel's secret atomic bomb program beginning in 1956 is drawn from numerous sources, chief among them author Seymour M. Hersh's excellent investigative

history, *The Sampson Option* (Random House: New York, 1991); *Nuclear Deterrence,* Shai Feldman (Columbia University Press: New York, 1982); *Six Days of War,* Michael B. Oren (Oxford University Press: Oxford, 2002); interview with Tom Moberly, engineer with TRW; reports of the Canadian Nuclear Association; Hamza, McKinnon.

30 "By 1971, Khidhir Hamza . . . in the brown flatlands of the Tigris." Hamza, McKinnon; *Saddam's Bombmaker; Le Monde,* Paris, p. 1, September 25, 1975; interview with Entifad Qanbar, a former Iraqi civil engineer who worked for Hussein's interior ministry. He defected to the United States in 1985.

40 "The two Israeli generals . . . 'we may have one or two ideas.'" Ivry, Perlmutter; *Flames over Tammuz,* Shlomo Nakdimon (Edanim Publishers: Jerusalem, 1986). A respected veteran Israeli journalist, Nakdimon was the first to reveal the deep political infighting over Osirak within Begin's cabinet, from the earliest days of mission planning until its final execution in 1981; *Gideon's Spies: Mossad's Secret Warriors,* Gordon Thomas (St. Martin's Press: New York, 1999).

MISSION IMPOSSIBLE

47 "The drivers of the two cargo trucks . . . Hofi and Mossad still had work to do." Many of the precise details of the sections dealing with Mossad's secret missions to derail the construction of Osirak come from Victor Ostrovsky and Claire Hoy's fascinating *By Way of Deception* (St. Martin's Press: New York, 1990). A former Mossad agent, Ostrovsky revealed many of the Israeli secret service's classified operations over three decades, outraging the spy agency. I was able to verify much of his account of the agency's Osirak exploits through other sources, who asked to remain anonymous. Also, Thomas, McKinnon, Hamza, Karsh & Rautsi, Ivry.

51 "In July 1979, just months . . . *nineteen* nuclear reactors for Saddam." Hamza, Karsh.

54 "Butrus Eben Halim was an unremarkable . . . Meshad's murder would go unsolved, if not unforgotten." Once again, I have drawn heavily from Ostrovsky, as well as from several other sources speaking on the condition of anonymity. The Mossad terms come from Thomas and Ostrovsky. Also, Hamza, Karsh, McKinnon.

64 **"The French, however, soon . . . French scientists were still in charge."** Hamza, Karsh, Nakdimon, McKinnon.

66 **"From the earliest days . . . After all, what were friends for?"** Technical details about the F-16 come from *Jane's Aircraft Upgrades,* Ninth Edition, David Baker (Jane's Information Group Limited, 2002) and interviews with USAF Lieutenant Colonel Jeff Dishart, as well as McKinnon, Ivry; interviews with mission squadron leader Zeev Raz and mission pilot Doobi Yaffe.

71 **"Conceived as a faster . . . perhaps, a little crazy."** *Jane's,* Oren, McKinnon, Ivry, Raz; interviews with mission backup pilot and present military attaché to the Israeli Embassy, Washington, D.C., Rani Falk.

73 **"In the fall of '79 . . . the second team to Hill."** Interview with second team leader Amir Nachumi; Ivry, Raz, Yaffe, Oren.

81 **"By February, the Italian . . . buying Ivry more time."** Interview with mission pilot Hagai Katz; Ivry, Karsh, Ostrovsky, Hamza.

THE WARRIORS

85 **"Hagai Katz couldn't believe his luck . . . What could Ivry be thinking?"** Interviews with mission pilot Relik Shafir and IAF F-16 pilot Dubi Ofer, one of the first twelve to go to Hill AFB; Falk, Yaffe, Raz.

95 **"The reports out of Baghdad . . . November was set as a tentative date."** Ivry, Hamza, McKinnon, Nakdimon, Perlmutter; IAF website.

100 **"Raz and his squadron . . . 'defending their own plants or destroying** *someone else's.'* **"** Ivry, McKinnon, Raz; *Washington Post* article, p. 1, by George C. Wilson, June 15, 1981.

104 **"Curiously, the NRC was not . . . Osirak would be hot by midsummer 1981."** Hersh.

107 **"Soon after Kivity and Saltovitz . . . God help them, probably nuclear."** Ivry, Raz, Nachumi, Katz; phone interview with mission pilot Ilan Ramon in Houston.

NOTES

THE WAITING

113 **"For months, Operation's engineers . . . And then things got complicated."** Ivry, Raz, Falk, Yaffe, Shafir, Katz, Nachumi; interviews with Raz's wingman, Amos Yadlin.

119 **"Ever since Ayatollah . . . only hope of completing the mission and
returning to base."** Hamza, *Saddam's Bombmaker*, Karsh & Rautsi, Ivry,
Nakdimon, Falk, Yaffe.

122 **"While the pilots practiced targeting . . . accomplished was taking
one another out?"** McKinnon, Dishart, Raz, Falk; Raz interview with operations
commander Aviem Sella for use in this book.

125 **"In January 1981 . . . not going to change anything."** Nachumi, Raz.

126 **"Several weeks later . . . Syria's new SAMs."** Yadlin.

126 **"Commander Spector was both . . . too late to turn back."** Oren,
Yadlin; interview with mission pilot Iftach Spector; Ivry, Nachumi, Raz, Falk.

133 **"By March 1981 . . . would do it anyway."** Ivry, Raz, Katz.

135 **"By the end of March . . . 'call it Operation Babylon.'"** Nakdimon,
Perlmutter, Ivry, Raz.

136 **"A week later, at Ramat David . . . 'what else matters?'"** Falk, Raz,
Yadlin, Nachumi.

WHEELS-UP

141 **"Heading back after the security . . . Ivry wondered."** Ivry, Perlmutter,
Nakdimon, Falk, Yadlin, Nachumi.

145 **"Begin read the note . . . Would this be the Sunday?"** Peres's letter is
reprinted in Perlmutter, McKinnon, and, in original Hebrew, in Nakdimon. The
political ramifications are recounted by Perlmutter and Nakdimon, who report-
edly had a source inside Begin's security cabinet.

147 **"The friction between the two . . . he alone was feeling now."** Raz.

149 **"On Wednesday, June 3 . . . dinner that night."** McKinnon, Ivry, Yaffe.

152 **"The question of whether . . . And that was that."** Raz; Yaffe and his wife, Michal; Katz; Yadlin.

154 **"The Friday before the mission . . . 'I hope I collect.'"** Yadlin, Yaffe, Spector, Shafir, McKinnon.

157 **"Sunlight from the first . . . 'God be with you.'"** Raz, Spector, Ivry, McKinnon, Sella, Yadlin, Nachumi, Yaffe. Eitan's speech was first reprinted by McKinnon in 1986.

164 **"The pilots suited up . . . he could have drawn."** Spector, Katz, Raz, Nachumi, Yaffe, Falk.

SIXTY SECONDS OVER BAGHDAD

172 **"The brown banks of . . . in their headsets: 'Grazen.'"** Raz, Ivry, Yadlin, Katz, Nachumi, Spector, Shafir. The checkpoint codes were given to me by Katz and Ivry.

182 **"Miles behind . . . lights from a small town."** Raz, Spector, Yadlin, Katz, McKinnon.

187 **"Khidhir Hamza stood outside . . . all Israel, it seemed, waited."** Hamza, Yadlin, Nachumi, Yaffe.

191 **"As Raz began to nose . . . any pursuing MiGs."** Raz, Yadlin.

194 **"One of the French electricians . . . the man exclaimed, astonished."** Hamza, McKinnon, Katz, Yaffe, Nachumi, Shafir. The descriptions of the bombing runs and the targeting are based in part on voice tapes from the pilots' helmet microphones and videotape from the F-16 nose cameras, which I was able to view. The quality of the tapes is quite good. Spector himself told me about blacking out, though he wasn't sure what had happened. He said he just missed. Nachumi was convinced Spector experienced blackout, a not-uncommon occurrence during high-G maneuvers.

CHECK SIX

202 **"Khidhir Hamza stood frozen . . . Everyone suddenly felt thirsty."** Hamza, Raz, Ivry, Nakdimon, Michal Yaffe, McKinnon, Yaffe, Katz.

210 **"Five hundred and eighty miles . . . across the installation."** Hamza; *Time,* June 15, 1981.

211 **"As darkness fell at Etzion . . . ever celebrate that holiday again."** Falk, Nachumi, Katz, Raz, Yadlin, Yaffe, Shafir, Ivry.

217 **"Richard V. Allen . . . 'terrific piece of bombing!'"** *President Reagan: The Role of a Lifetime,* Lou Cannon (Touchstone/Simon & Schuster: New York, 1991); Hersh; interview with Caspar Weinberger; interview with Richard Perle.

221 **"Monday morning, July 8 . . . 'which is near Baghdad. . . .'"** Katz, Raz, McKinnon.

224 **"That Sunday, the cavernous . . . deliver within the month."** Thomas.

EPILOGUE: BLOWBACK

225 **"Monday morning, June 8 . . . 'start from scratch.'"** Hamza, Karsh.

227 **"The storm of indignation . . . was quietly resumed."** McKinnon, Perlmutter, Hersh; IDF website; *New York Times,* editorial page, June 9, 1981; *Time,* cover story, June 15, 1981; Cannon, Ivry.

231 **"Perhaps in somewhat . . . whom Hofi was referring to."** The anecdote about Begin accidentally referring to Dimona comes from Hersh.

233 **"The Arab nations . . . chemical weapons programs."** Thomas, Hamza.

236 **"On June 7, 2001 . . . signed: 'Dick Cheney.'"** Falk, Katz, Raz, Nachumi, Ivry.

INDEX

A

Abul-Khail, Abdullah, 21
Admoni, Nahum, 233
aircraft, 68–73. *See also* F-15 aircraft;
    F-16 aircraft
Al Auja, 11, 13
Allen, Richard V., 217–18, 220
al-Tuwaitha. *See* Nuclear Research
    Center of Atomic Energy
AMAN, 54, 55, 58
American Enterprise Institute, 239
Arab Legion, 15
Arab Rejectionist Front, 42
Arab states, reactions to air strike,
    228, 229–30, 233
Arafat, Yasser, 21, 22, 142, 231
Arbel, David, 55, 58
Argentina, reaction to Osirak air
    strike, 228–29
atomic bomb. *See* nuclear weapons
Atomic Energy (AE), Iraq, 34, 95,
    234; Hamza report to, 30–31,
    33–34. *See also* Nuclear Research
    Center of Atomic Energy
Attia, Ali, 20, 34
Aziz, Tariq, 51

B

Babylonian captivity, 12
Baghdad: in 1950s, 14; Palestinian
    refugees in (1960s), 21–22
Bahrain, reaction to air strike, 230
Bahr al Milh Lake, 182–83
Baker, James A., 219
Bakr, Ahmed Hassan al-, 18, 19, 20,
    51
Ba'th Party, 14–15, 51; Qassem assas-

sination attempt, 16–17; Qassem
    ouster, 18–19
Ba'th Revolutionary Command
    Council (RCC), 19, 21, 51
Begin, Menachem, 59, 105, 142, 152,
    212; during mission, 177, 182,
    190, 205, 210; and foreign reac-
    tions to air strike, 229, 230; me-
    dia apprisal of air strike, 222–24,
    232; mission cancellations, 120,
    145–47; mission consideration
    and approvals, 41–44, 97–100,
    135–36, 147; notifies U.S. of air
    strike, 222; political aftermath,
    230–33
Ben-Amitay, Udi, 94, 126
Ben-Gurion, David, 23, 24, 26, 152
Ben-Nun, Avihu, 69–70, 118, 119
Bergmann, Ernst, 24, 30
Biran, David, 43
Black September, 22
bombs, for Osirak air strike, 101–2,
    103–4, 120–21, 134, 139
Brazil, nuclear negotiations with Iraq,
    54
Britain: KH-11 intelligence access,
    105, 228; reaction to air strike,
    228; in Suez War, 24–25; view of
    Iraqi reactor purchase, 41–42
Brown Boveri, 64
Bush, George H. W., 236

C

Cairo, Saddam Hussein in, 17–18
Camp David Accords, 3, 142
Camp David summit, 105
Carter, Jimmy, 68, 70, 81, 105

ABOUT THE AUTHOR

RODGER W. CLAIRE, a former magazine editor, is the first journalist to have been granted complete access to all of the individuals involved in the raid on Osirak and to classified materials detailing it. The author of numerous articles and two screenplays, he lives in Los Angeles.